Economics in Horticulture

Other Related Books

Ayurvedic Medicinal Plants of India, Vol. 1-2 (Set)	Bhutiya, R.K.
Bacterial and Viral Disease and their Management in Horticultural Crops	Reddy, P.P.
Bamboo Flowering and Rodent Outbreaks	Sridhara, S.
Biofuels : Potential and Challenges	Pandey, A.K.
Biofumigation and Solarization for Management of Soil-Borne Plant Pathogens	Reddy, P.P.
Biointensive Integrated Pest Management in Horticultural Ecosystems	Reddy, P.P.
Chickpea (Gram) Status and Cultivation Technology	Shekhawat, S.S.
Controversial Herbal Drugs of Ayurveda	Saroya, A.S.
Diseases of Horticultural Crops: Nematode Problems and their Management	Reddy, P.P.
Flavouring and Fragrant Resources of India	Sood, S.K.
Flowers and Vegetables of India	Sud, R.K.
Folk Herbal Medicine and Drug Discovery	Katewa, S.S.
Fungal Diseases and their Management in Horticultural Crops	Reddy, P.P.
Hort-Agro Project Reports	Sud, R.K.
Horticulture and Livelihood Security	Nath, P.
Insect, Mite and Vertebrate Pests and their Management in Horticultural Crops	Reddy, P.P.
Integrated Nematode Management in Horticultural Crops	Reddy, P.P.
Ornamental Plants for Gardening	Chopra, V.L.
Plant Protection in Horticulture Vol. 1-3 (Set)	Reddy, P.P.
Productivity Enhancing Technologies for Horticultural Crops	Reddy, P.P.
Recent Trends in Disease Management of Fruits and Seeds	Prasad, M.M.
Salinity Tolerance in Plants: Methods, Mechanisms and Management	Garg, B.K.
Science of Tea Technology	Ahuja, P.S.
Silvipasture in India: Present Perspectives and Challenges Ahead	Palsaniya, D.R.
Tropical Forest Ecosystems Structure and Function	Singh, V.P.
Tropical Planting and Gardening 6th Edition	Barlow, H.S.
Useful Herbs of Planet Earth	Daniel, M.
Vegetable Gums and Resins	Howes, F.N.
Weed Seed Biology	Vanangamudi, K.

Economics in Horticulture

M.J. Sargent
Lecturer in Horticulture
University of Bath

United Book Prints

Reprinted in India by:

UNITED BOOK PRINTS
(Imprint of Scientific Publishers, INDIA)
4806/24, Ansari Road, Daryaganj,
New Delhi 110 002, INDIA
E-mail: info@scientificpub.com
Website : www.scientificpubonline.com

Print : 2023

ISBN: 978-93-83692-02-6
ISBN: 978-93-87741-81-2

Printed in India

CONTENTS

ACKNOWLEDGEMENTS

This book is the product of experience with growers, economists and more recently with university students of horticulture. Their stimulation and encouragement is gratefully acknowledged.

I am particularly indebted to Professor Leonard Broadbent, whose helpful suggestions and skilful editing, have greatly improved the readability of the final product. I would also like to thank my other colleagues at Bath for their encouragement and helpful comments on early drafts.

The assistance is also gratefully acknowledged of workers and organisations who have allowed me to reproduce tabular material to give the text greater realism. In particular the Publicity Department of the Francis Nicholls Organisation, Smithfield Market, Birmingham, who provided the frontispiece. The permission of the Controller of Her Majesty's Stationery Office has been obtained for the use of certain tables.

Finally, I am deeply indebted to my wife, Valerie, who not only had to endure many lonely evenings during the preparation of the book but has also helped by reading early drafts and proofs.

INTRODUCTION

The word economy has two Greek roots referring to household or estate management. Economics has been generally in use only since the early part of this century. In more recent years it has broadened in scope and today encompasses many specialist disciplines. Economics is concerned with the exchange of goods and services between people and the organisation of scarce resources. This definition clearly needs elaboration.

Returning to the older household management root can provide us with a very simplified example. The urban housewife of today has a limited income to spend organising her budget so that the family is fed, clothed and maintained. She allocates her scarce resources between these competitive uses of her money to gain the greatest satisfaction possible. Food and heating may well have priorities and thereafter she may have a scale of preferences towards the more luxury types of consumption such as recreation and holidays. In just the same way as she allocates her scarce resources so does the Government in organising its annual budget and so does the horticultural producer in organising his spending. Our housewife may gain satisfaction from her spending by the sight of a happy contented family. Benefits accruing from Government expenditure are much more intangible and difficult to quantify. The grower has more positive benefits in terms of moneys coming from the sale of produce in a market. This is his source of moneys for further expenditure and his own family maintenance. The Government may raise its moneys by taxation on land, imports and incomes of the population. Our housewife's income will have been derived from her own or her husband's work. In reality of course the husband might be employed by the Government as a civil servant. He may pay taxes to the Government on his income. His wife may buy produce from the grower, who in turn is also a consumer and taxpayer, but in addition may well receive assistance from Government expenditure.

Exchange

In our example we can identify logically and conveniently distinct parts of the exchange economy. The grower is concerned with production of crops using the resources of land, capital and labour. In the exchange economy he sends his produce to consumers, many living in urban centres geographically removed from his holding. The process of transferring products from the point of production to the consumer is called *marketing.* The grower's organisation of production and marketing, usually with profits as a goal, is called *management.* Central Government concerned with the smooth running and strength of the national economy may need to influence this exchange system to improve its overall efficiency. It may for example introduce a wage restraint or income tax in order to manipulate consumers' incomes; the grower may receive grant aid or tax concessions and by these means a national *policy* is adopted towards a particular economic end. Stimulation of exports, retarding of imports, expansion of particular forms of production or consumption may be chosen. The professional economist is concerned with the efficient use of resources at individual, business and national level; as an observer of consumer and producer activities and of the effects of Government policies he carries out an *economic analysis.* He may construct simple graphic or more complex mathematical models of particular situations from which he tries to understand past situations and to predict with reasonable certainty future results from given actions.

Money is the basis and common exchange medium in our example and it is in markets that this exchange occurs. It is here that consumers make their desires known and values are determined for the goods or services exchanged. Even in our very simple example there are clearly many more people involved than the three central figures. The whole population of the country and if international trading is included, of the world, are involved in economic exchange. Man is a social animal and society comprises a wide genetic spectrum of races and of behavioural pattern to given stimuli. Economics is concerned with the behaviour of people in allocating their scarce resources – it is a social science.

Economics and social sciences

The social sciences include such disciplines as sociology, anthropology, geography and psychology; subjects less familiar to the student of horticulture who more commonly finds himself associated with the natural sciences. This is particularly true of the horticultural colleges and university departments. In common with all sciences measurement features prominently in social investigations. The objects concerned in this measurement and the units used are perhaps the earliest stumbling block encountered by the student new to the subject. Once these new concepts have been mastered there are some similarities of approach to other sciences and much is common sense. One cannot do better at this stage than to quote the well known economist Colin Clark who in a recent book called *Starvation and Plenty* says:–

'So much of economics consists of statements which are obvious as soon as one comes to look at them but which for various reasons seem to escape the notice of active statesmen and of learned professors.'

The economist in particular does face some rather different problems. It is very difficult for him to carry out a controlled experiment in the same manner as one might for example investigate the effects of temperature on photosynthesis in plant physiology. He relies on observations, theories and conclusions drawn from statistical records of the past. Value judgements inevitably feature prominently. The student must learn early on to distinguish between fact and opinion and appraise them in relation to the current state of knowledge. As a young subject its theories are not highly developed and capable of proof; mathematical tools as aids to the expression of new concepts are a comparatively recent addition.

A student new to economics will soon discover that it is an opinionated subject. Everyone knows something about money, trade, taxation and will offer advice to the exchequer on the running of the economy. Even professional economists disagree on their methods of analysis and interpretations of situations. There is an old joke, often attributed to Keynes, that if you take any six economists you have seven opinions. Inevitably perhaps there may commonly be a large gap between theoretical analysis and reality. Having to live through the implementation of a treatment prescribed by an economist has been an unhappy experience for more than one individual.

Agricultural economics and horticulture

Agricultural economics has evolved as a separate study largely within this century. This has probably come about because its origins were often in institutions concerned with husbandry aspects of agriculture. The study of the business management aspects of these husbandries has gradually led to deeper investigations of the economics of agriculture. Today agricultural economics has its own learned societies nationally and internationally and is an established discipline in universities.

Studies of the economics of horticulture have always been regarded in the past as simply part of agricultural economics. Often horticulture has found itself something of a 'cinderella' in this respect. The producers of goods for relatively luxury consumption, such as flowers, fruits and vegetables, have not commanded the same attention as producers of bulk carbohydrates and more recently animal proteins. Not surprisingly this is reflected in a dearth of statistical and documented information about many horticultural crops and situations. In the past twenty years however there has been some improvement in the situation and one of the most encouraging signs has been the establishment of a clearly viable sub-group of the International Society for Horticultural Science to study horticultural economics.

Horticulture defined

In lecturing to his own students the author defines horticulture as that part of agriculture concerned with the production of relatively high value crops, not including those for bulk carbohydrate production or the extraction of chemicals or for animal production. This is a negative definition in that it delineates a residue of crops relevant to United Kingdom horticulture. In another climate other crops may be regarded as horticultural and others seen as simply part of agriculture. Even in the UK vegetable crops and even nursery stock are being grown to an increasing extent on large farms having agricultural crops as well in terms of the definition above. Fruit growing has always featured in mixed farm cropping. In these situations the economist cannot escape consideration of agricultural crops if he is investigating the business economy.

A further problem of the use of the word horticulture is its equation in the minds of many of the general public with gardening. Horticulture is a sophisticated technology and education of the general public to this effect is still necessary. It is a professional and commercial, not amateur, occupation and in this book as the title implies we are concerned solely with commercial horticulture.

Even commercial horticulture today can be sub-divided into specialist disciplines. The two broad categories of crop production and amenity (including nursery stock) are usually recognised. The latter category is different in that it is concerned with creating an effect. A display or amenity planting is only achieved at cost and it may even help to 'sell' a resort or town to tourists but it is not sold itself. Nursery stock production is a unique area of horticulture as well. Although its products are marketed they are not sold in the same manner as most fruit, vegetables and flowers. Most nursery stock producers are able to determine the price of their products themselves. This does not mean that they can charge what they like because they are still in competition with other nursery stock producers. In this respect they are more akin to producers in other non-agricultural industries.

This book is concerned with economics in commercial crop production. It may be that we need a new word to describe this type of horticulture but any new word would have to achieve international acceptance.

Economics in horticulture

Horticultural technology is related to many other disciplines of both the arts and sciences. Economics with its roots in the social sciences broadens its outlook considerably for it is truly inter-disciplinary and international. Furthermore the comparative youth of horticultural economics as a study means that it is still possible to achieve originality from seemingly quite elementary investigations.

The writer of a textbook cannot hope to mirror all the opinions of economists in an introductory text. It is intended here to lift the veil surrounding economics in horticulture. It is hoped that the reader or student or grower or adviser or even other economists might find encouragement to delve further into the literature of general economics. It can only be beneficial for horticulturists to become more familiar with the social sciences and for them to try to understand the particular characteristics of the industry. A reference list for suggested further reading is included at the end of this book. It must be stressed however that economics is a dynamic science. The economic situation is undergoing continual change. The serious student should try to relate daily news bulletins and their contents to his knowledge of economics.

Ideally this broad spectrum subject ought to be read with the chapters in parallel. Inevitably however one chapter has to follow another: the cyclic arrangement of chapter headings seemed to the author the next best solution. Thus we start with the overall national economy and conclude at national level again in considering governmental policies. In between we gradually focus down on the individual grower and his business.

CHAPTER 1

THE NATIONAL ECONOMY

The value of all goods and services produced in a country by the resident population in a year is called the Gross National Product (GNP). It represents the result of using resources and managerial control in production. It is the wealth produced by the people. As the country becomes more wealthy the size of the GNP increases; this is called economic growth.

1.1 THE GROSS NATIONAL PRODUCT

The term GNP is an accounting concept but accounting procedures are fraught with complications of nomenclature, composition, allocation and interpretation. Accountants may not agree on how to value or describe a particular feature not only within the country but internationally too. The central problem is one of valuation of the contribution made by each member of society to the total. It is relatively easy to measure the output of say coal-miners or horticultural workers but how do you put a realistic value to the output of firemen, university lecturers, students or housewives? In practice they are generally valued at cost, that is their salary or cost to maintain. A value for housewives, who undoubtedly work very hard, is often omitted altogether.

These limitations may be important particularly when making international comparisons of economic development. The values may be further affected by a change in social structure, for example more wives might go out to work, or more households engage servants, in one country than in another. Furthermore it is important that each item should not be accounted for more than once. If fertiliser has been bought and used in the production of carrots and the value of the fertiliser included in the value of the carrots it will have been double-counted. To avoid double counting we often consider only the *value added* to a product by a particular business. To do this we deduct from the total value of its output the cost of all items such as fuel, containers, materials and services used in its production. Thus the value added by a business is the net contribution made by the business to the national output. After paying taxes to the Government and allowing some funds for depreciation, the remainder becomes the incomes of the employees and owners of the organisation. The total output of the country measured in this way is known as the Gross National Product (GNP) at *market prices.* Sometimes the part of this sum due to the Government as 'indirect' taxes is deducted, and the resultant total is known as the Gross National Product (GNP) at *factor prices,* which is the cost of employing the factors of production.

As time passes, production, plant and machinery decline in value as they obsolesce; this process is called *depreciation.* To maintain business output this must be made good by replacement or repair. For some purposes the value of this depreciation is deducted from the GNP and the resultant is known as the Net National Product at market or factor prices depending on whether taxes are deducted or not.

The national income may be measured in different ways and it is in the United Kingdom (UK). It may be measured by adding together the incomes of all the inhabitants of the country or alternatively measuring the total spent on consumer goods and services together with the value of investments made. Both of these methods are not without complications and rules of compilation that must be complied with in order to

Table 1.1

Gross National Product by industry (UK)

	£M	
	1959	1969
Agriculture, forestry and fishing	877	1197
Mining and quarrying	689	678
Manufacturing	7509	13346
Construction	1236	2559
Gas, electricity and water	571	1364
Transport	1293	2348
Communication	393	930
Distributive trades	2570	4193
Insurance, banking and finance (including real estate)	625	1250
Ownership of dwellings	835	1991
Public administration and defence	1262	2391
Public health and educational services	822	1987
Other services	2340	5073
Less stock appreciation	−90	−815
Residual error	230	−342
Gross domestic product at factor cost	21 162	38 150
Net property income from abroad	260	451
Gross National Product at factor cost	21 422	38 601

standardise the accounts. By accounting in several ways, an inbuilt check is possible in that the totals should agree. The GNP only need concern us here but for further discussion of these methods the student is referred to introductory texts in economics.

The GNP account for the UK for 1969 taken from the 1970 HMSO Blue Book *National Income and Expenditure* is reproduced as an example in table 1 and is compared with that of 1959. Not all countries have practicable methods of estimating the national product in such detail. The only feasible system in some countries may well be to merely total output.

A further cautionary note must be sounded on the subject of international comparisons between levels of GNP. After ascertaining that the accounts have been compiled in the same manner we should divide the total by the population to arrive at a figure of GNP per head. Further complications of comparison arise from differences in ways of life and prevailing price levels. The important feature to grasp is that all countries are at *a* stage of economic growth and have a particular GNP at a moment in time.

1.2 AGRICULTURE AND ECONOMIC GROWTH

The countries of the world have different levels of GNP but all have evolved from primitive origins. Some countries like those of North America and Western Europe have travelled a considerable distance along the path of economic growth and development whilst others like Afghanistan, for example, are still at an early stage. It is now generally agreed that agricultural man dates from about 7000 BC. Through the following 8500 - 9000 years the development of different types of agriculture can be traced in distinct centres. So called *shifting agriculture* was a common form of early crop production: the land is cleared, often by burning, tilled until its fertility is depleted and then abandoned for a new site. In remote parts of Sweden the technique persisted until the 1920s and it can still be found in parts of Africa, the East Indies and South America.

In the very primitive agricultural communities the population is often tribal and certainly family and clan connections are strong. Furthermore these families are mainly concerned with survival and are generally self-sufficient, technology making very little impact in their lives. This is the so called *subsistence economy* and probably the whole pre-Newtonian world was of this character including the Dynasties of China, The Middle East and Mediterranean civilisations and Medieval Europe. These, you may argue, were highly developed civilisations. However, the political power and wealth was vested in a very small proportion of the population and the rulers were often tyrannical. Agriculture in these traditional societies was the prime industry, being the main occupation and source of income.

Transition

To change from this traditional state and to accelerate economic growth, changes must clearly occur in society. This phase of development is often called *Transition* and during it a market structure emerges initially very simple in concept but essential for the exchange of goods and services. The markets often arise at centres of population and communication which become towns and cities. The first alternative industries to agriculture, although often closely related, also commonly arise at these centres; for example processing factories, forestry product manufacturers and fertiliser plants. The new and usually higher paid employment offered by these new industries attracts rural workers to the urban centres. Concurrent with this development institutions such as banks, finance houses, legal systems and local government organisation begin to evolve. Education facilities begin to expand and to embrace a greater proportion of the population. The people begin to think and develop a nationalistic spirit. Common to all these developments is the need for publicly available funds so some form of tax collection emerges.

A crucial feature of this phase is the breaking of the so-called poverty cycle which in simple terms takes the following form. Agriculture starts in a low state of technology because of the country's stage of development which means that low or no incomes are generated by the farmers. This low income means that inevitably there is low investment in such things as new cultivators, fertilisers, crop protection materials and tools, so that agriculture is maintained in a low state of technology and the pernicious and difficult-to-break cycle continues. It is in the transitional phase of economic development that agriculture carries an immense burden in the drive for economic growth. Agriculture is required to

(1) *produce sufficient food to*

(a) feed the population on the land

(b) feed those living in and moving to the towns

and

(c) satisfy the increase in demand for food from the larger total population with its improving income and resultant purchasing power.

(2) *produce capital for*

(a) new investment in other industries

and

(b) establishment and improvement of transport and communication networks. In both cases capital is raised by taxation or exportation.

(3) *release labour to move to the urban centres*

That this is a heavy burden to bear is plain to see. Agricultural progress is truly a prerequisite of industrial progress. The temptation to try to move too quickly is great. In these situations agriculture may get out of phase and the country may well find itself unable to feed its population and to generate sufficient capital for further development. The importance of careful planning cannot be over-stressed and many countries (including so-called 'developed' ones) do produce regularly five and ten year plans to try to ensure coordination of activities towards balanced growth. Much technical and financial aid is directed by developed countries towards this important transitional phase in the developing countries.

Horticulture at this stage has been relatively unimportant mainly because it is concerned with products associated with higher incomes and more sophisticated diets. Particularly in the past twenty years, however, horticulture has emerged even in the developing countries. Often intensive cropping systems are easily added to existing agricultural systems and are used to help raise important foreign exchange earnings. This trade itself depends on other countries having developed far enough to become importers of this 'luxury' produce. Many African, South American, Mediterranean and Far Eastern countries now export horticultural products and even processed foods by sea and air to European and North American markets. The importance of horticulture in the economy of these countries is increased also with attempts to improve the diet and health of rural populations by introducing vegetable and fruit culture into subsistence farming. This has been done with some success in Africa against considerable resistance, although in West Africa passing ships have been serviced with fresh produce for over 100 years.

As trade and economic activity increase through the transitional phase there comes a point when economic growth becomes automatic, although not necessarily at a satisfactory level. Put another way – the country's economy overcomes resistances to development. Economists refer to this point as the 'take-off'. It is not an identifiable moment such as when aircraft 'take-off' and there are many ways of achieving it. It is only discernable in retrospect and often can be associated with a particular stimulus, for example in Germany the revolution of 1848; in India attaining independence in 1947; in Denmark a switch from meat to milk production in the late nineteenth century.

Having achieved 'take-off', problems are by no means in the past. As economic development progresses there arise problems of maintaining growth and balance between sectors of the economy. It may be difficult at this stage to keep long-term aims in view. Some economists have suggested that maturity is reached when society can demonstrate a capacity to exist beyond a structure comprised of solely primary industries. The Food and Agriculture Organisation (FAO) of United Nations (UN) in 1952 divided countries into developed and developing categories, all those having equal living standards to those of Western Europe and North America being regarded as developed. Even in these countries there is considerable variation in stage of maturity or development.

Economic 'maturity'

During the maturing phase the main emphasis is still on the maintenance of a balanced role for agriculture. Two dominant problems that may arise are firstly, fluctuating world market prices may be damaging to a particular country's economy. Countries that have too large a share of their economy dependent upon one crop (mono-cropping) find that the whole economy fluctuates in fortune in unison with the crop itself. Brazilian coffee and Mauritian sugar are examples of such crops. Secondly, the birth rate may increase as food becomes more plentiful to a point where it becomes difficult to meet the increase in demand for food. Imports of food become necessary at the expense of valuable exchange currency.

In practice it is not always easy to maintain satisfactory growth rates, a fact well-known to many developed countries in the past decade including the UK. It is very tempting for countries to try to miss essential stages and coordinated planning is still imperative. Horticulture becomes more strongly defined through this period and home markets in particular expand as incomes rise and the consequential spending ability of the population increases.

At maturity or as it is called by many the 'age of high consumption' six main economic characteristics often feature prominently.

(1) There has often been a pursuit of external power and influence and even colonisation on the part of nations passing through this phase.

(2) Governmental powers and organisations are used to redistribute wealth so that it is shared *more* generally by society.

(3) Agriculture declines sharply in importance as an occupation. (Illustrated in table 1.2)

(4) Diets become more sophisticated and consumers are prepared to purchase a service, with their foods partially or wholly processed and often ready for consumption. These are called convenience foods.

(5) Consumption of food expands well beyond basic requirements and the proportion of consumers' incomes spent on food declines (table 1.3).

(6) Leisure time relative to working time becomes more plentiful.

Most of Northern Europe, the United States of America, Canada and Australasia have reached this stage. The UK was the first country to pass 'take-off' at the end of the 18th century and now only produces about half of its food, importing the remainder; less than 3½ per cent of its population now works on the land.

Table 1.2

Agricultural employment as a percentage of total employment 1900-1962

	USA	*Sweden*	*UK*	*Italy*
1900	38	43	8	59
1910	31	41	8	56
1920	27	35	7	56
1930	21	30	6	47
1940	17	27	n.a.†	n.a.†
1950	12	n.a.†	5	43
1962	8	13	4	28

OECD (1965), 'Agriculture and Economic Growth'.

† n.a. = not available.

There is not one model pathway of economic growth. In this chapter we have considered only a generalised pattern. Each country has its own

Table 1.3

Food expenditure as a percentage of total consumers' expenditure 1950-60
(Calculated on the basis of current market prices)

	1950	1955	1960
Austria	48.9	45.9	42.2
Canada	24.8	23.1	22.9
Denmark	28.4	27.6	24.2
France	39.4	35.3	31.7
Greece	51.3	49.0	45.2
Netherlands	32.7	33.7	31.0
UK	28.4	31.4	28.8
USA	24.3	23.2	21.4

OECD (1965), 'Agriculture and Economic Growth'.

particular characteristics, problems and natural attributes. Australia for example is well endowed with minerals and has a favourable climate for agriculture so it has retained a high proportion of primary production in its economy even into maturity. Sometimes barriers to development can arise from such sources as wars, religious principles and land tenure traditions. Our general model is most valid for non-communist situations although obvious similarities in aims can be observed even in communist countries. Marx talks of feudalism, bourgeois capitalism, socialism and communism as being the stages of economic growth.

The dominant feature of all economic development for our purpose is the decline in importance of general agriculture in the economy and the accompanying progressive increase in importance of horticulture as the wealth of the country increases. The expansion of horticulture is directly related to the increase in disposable incomes of the population. Table 1.4 illustrates the importance of certain horticultural crops in mature countries and in some which are expanding horticulture to earn foreign exchange. The table is arranged in order of bulk carbohydrate (mainly cereals) production. This is slightly misleading in that Jamaica and Mauritius at the bottom of the table are large-scale sugar producers. Other important agricultural enterprises not included are plant fibres, oils and milk production.

1.3 CASE STUDY ILLUSTRATIONS OF HORTICULTURE IN THE NATIONAL ECONOMY

To illustrate this complex and very individualistic process of economic growth four case studies are presented. These have been chosen to illustrate the emergence and importance of horticulture within the economy of countries in a range of different situations. The UK situation is examined in greatest depth because this is possibly the best documented

Table 1.4

Gross agricultural production by commodity groups for selected countries

Country	Contribution to total agriculture in 1968 (%) Cereals	Meat	Fruit	Citrus	Vegetables
Zambia	58	13	–	–	–
India	45	3	4	1	5
USSR	33	17	1	–	3
USA	27	32	3	1	4
Taiwan	27	32	9	2	5
UK	21	36	1	–	4
Cyprus	10	18	28	15	8
Colombia	10	17	11	–	6
Netherlands	7	37	2	–	5
Israel	5	17	29	26	8
New Zealand	3	29	2	–	2
Jamaica	–	16	12	3	2
Mauritius	–	1	1	–	3

Adapted from FAO (1971). *Monthly bull. Ag. Econs and Stats* 20, 8.

and is longest in time past 'take-off'. New Zealand is an example of a mature economy where agriculture has remained very important in the economy. Taiwan and Israel illustrate, from different origins, the methodology and importance of exports in earning foreign exchange.

Case A. The United Kingdom (UK)
'Take-off' in the UK came at the end of the eighteenth century. It is of particular interest because not only was it the first country to attain this but the manner of accomplishment was unique. It came during the industrial revolution when rapid progress was made in the development of coal, iron, steam power and in particular textile manufacturing. These developments were followed and accelerated by the emergence of low cost transport systems. A large industrial structure was developed with considerable export trade, reaching 60 per cent of output by 1820 and depending greatly upon the importation of raw materials for its success. In fact the rate of increase of imports of raw cotton between 1781 and 1791, some 320 per cent, reflects plainly the rapid increase in economic growth at that time.

Agriculture and horticulture
We are fortunate that there has been documentation of the UK economy for many years starting early in the 19th century. The decline of agriculture as a contributor to the national product can be seen in table 1.5. During these 155 years the national product grew from about £230M in 1801 to £2000M by 1900, reaching £19 000M by 1955. In table 1.1 we

saw a more recent total of £38 600M for 1969 and the way in which this is shared between the different sectors of the economy. The agricultural sector including fisheries and forestry now only accounts for 3.1 per cent of the GNP.

Table 1.5

The structure of the British national product
(as percentages of total national income)

	Agriculture, forestry and fishing %	*Manufactures, mining and building* %	*Trade, transport and overseas incomes* %	*Government, domestic and other services* %	*Housing* %
1801	32.5	23.4	17.4	21.3	5.3
1851	20.3	34.3	20.7	18.4	8.1
1901	6.1	40.2	29.8	15.5	8.2
1955	4.7	48.1	24.9	19.2	3.2

Dean, P. and Cole, W. A. (1969). *British Economic Growth 1688-1959.* Cambridge University Press 2nd edn.

Horticulture as a commercial venture had its origins in cottage gardens and smallholdings. The stimulation was provided by the population expansion in the 16th and 17th centuries. Small quantities of cabbage, turnip, cauliflower, carrot, parsnip and some fruits had been sent to London markets since the reign of Elizabeth I. Some imports from abroad into London were also made at that time.

The hundred years between 1650 and 1750 brought the main development of the London market which in turn gave further stimulus to English agriculture. Horticulture in particular responded to this stimulus. Hop and vegetable growing became businesses in their own right, within easy reach of the City instead of being insignificant parts of general agriculture. The orchards of Kent and Essex too, came into their own. Some wholesalers who resold part of their country purchases to retailers also emerged, apparently not always being welcome as 'middlemen'.

During the 19th century the development of horticulture accelerated because of increasing incomes of the working population now able to afford to buy produce from organised markets. In 1843 the 200-year old Covent Garden market was covered; perhaps this is the best-known market in the world. Markets developed at other centres such as Birmingham, Manchester, Liverpool and Bristol, these becoming firmly established in the latter part of the 19th century. The railway system of the country also developed in this period bringing with it the ability to transport produce quickly and at that time cheaply from new areas of production to centres of distribution and consumption. By the end of the 19th century the main areas of production and market sites were established and they have changed little to this day.

The structure of agriculture and horticulture

UK agriculture is today orientated about animal production. Much of its effort is devoted to the maintenance of animals through the winter months

Table 1.6

Farm sales for 1969-70 in the UK
Total = £2114¼M

	%
Beef and veal	15.7
Pork and bacon	11.3
Mutton and lamb	4.1
Milk and milk products	21.7
Eggs	9.4
Poultry	5.7
Grain	11.1
Other farm crops	7.9
HORTICULTURE	10.7
Other	2.4

HMSO (1970). 'Cmnd 4321: Annual Review and Determination of Guarantees'.

Table 1.7

Value of UK output at the farm gate

	1964-65 £M	1966-67 £M
Vegetables in the open (a)	60.4	67.4
Fruit in the open (b)	50.5	46.6
All crops under glass (c)	43.3	46.4
Hardy nursery stock	14.3	15.9
Flowers and bulbs in the open	6.1	6.1
TOTAL	174.6	182.4

(a) Excludes peas harvested dry and for canning and freezing. The value of these was £9.7M in 1964-65 and £10.1M in 1966-67. Potatoes also excluded.
(b) Subject to adjustment on final analysis of 1966 Orchard Fruit Census.
(c) Includes mushrooms which were valued at £7.9M in 1964-65 and £8.7M in 1966-67.

HMSO (1967). 'Examination of the Horticultural Industry'.

when due to the climatic pattern outdoor grazing is poor. Grain is used for feeding together with brassica and other crops grown specially as feeds. Table 1.6 shows the structure of agriculture by value of total sales. The dominance of animal production is very evident and horticulture measured in this way represents just over 10 per cent of total agriculture. This puts horticulture in perspective as only 0.1 per cent of the national product. However if horticulture is measured in terms of its value added it appears to use its labour and capital more efficiently than agriculture generally. A proportion between 16 and 20 per cent of agriculture has been suggested by Dr R.R.W. Folley of the University of London as being more representative.

Although it covers a period of only four years, table 1.7 gives a more detailed insight into the contribution of particular divisions within commercial horticulture. Flowers and other luxury crops cover about one quarter of production.

Agriculture as a whole produces approximately one half of the food consumed in the UK, the remainder being imported. In horticulture too, the nation is not self-sufficient as is apparent from table 1.8. Although horticulture has expanded in recent years and despite an apparently high proportion of supplies met by home production, UK growers have actually lost ground in the home market. In the decade 1955–65 home vegetable and fruit production declined by 6 per cent and 14 per cent respectively. Flower growers did increase their share but only by 1 per cent.

Imports account for 38 per cent of supplies and of these one quarter come from the EEC, 13 per cent from the Channel Islands, other commonwealth countries contribute 22 per cent and all other countries 40 per cent. Exports from the UK are insignificant, amounting to a few exploratory sendings, but interest in expanding this trade is increasing.

Case B. New Zealand

New Zealand's short history has been strongly influenced by its dependence on overseas trade for development and economic progress. It is also of great interest because agriculture has remained a major sector of the economy whilst a good standard of living and large GNP have been achieved. (table 1.9).

After the foundation of agricultural industries in New Zealand over 120 years ago early trade was principally with Australia. With the advent of refrigerated shipping there followed a spectacular change in trade and economic growth. Exports jumped from £6.3M in 1880 to £13M in 1900; they doubled again by 1920 and in 1965 stood at £375.5M. Imports have kept pace and New Zealand is traditionally now in deficit on its invisible account.

Australia is still the closest overseas market (1200 miles) but the main markets of Western Europe are 12 000 miles away so transport costs are inevitably high. There is a high dependence upon primary products in this

Table 1.8

Total supplies of horticultural produce by type, 1965-66

Type of production	*Home production* £M	%	*Imports* £M	*Total supplies* £M	*Proportion home grown* %
Outdoor vegetables	66.8	37.2	18.8	85.6	78.0
Glasshouse vegetables	26.8	14.9	39.7	66.5	40.3
Glasshouse flowers	18.3	10.2	0.8	19.1	95.8
Total glasshouse	45.1	25.1	40.5	85.6	52.7
Outdoor flowers	21.4	11.9	13.3	34.7	61.7
Total glasshouse and flowers	66.5	37.0	53.8	120.3	55.3
Fruit	46.3	25.8	39.0	85.3	54.3
TOTAL HORTICULTURAL OUTPUT	179.6	100.0	111.6	291.2	61.7

Includes Channel Islands £12.3M

Hinton, W. L. (1968). 'Outlook for Horticulture'. *Occ. Paper* No. 12. Farm Econs. Branch, Cambridge University Press.

trade; 90 per cent of exports are derived from sheep and cattle and Britain is still the recipient of over 90 per cent of New Zealand's butter, cheese and lamb. These primary products are very vulnerable to world price fluctuations. Constant searching goes on for import substitution and

Table 1.9

Horticulture in the New Zealand Economy

	1963	1964	1965	1966	1967	1968
GNP $M	2921	3196	3487	3744	3911	4032
Gross farm income $M	654	764	792	853	824	826
Horticultural income $M	57.1	62.4	69.6	76.0	79.9	83.4
Horticulture as a proportion of agriculture %	8.7	8.2	8.9	8.9	9.7	10.1

Adapted from Johnson, R. W. M. (1970). *Bull.* 8. *Lincoln Coll.* N.Z.
Ed. Morrison, T. M. 'Econs. and Management of Vegetable Production'. 10-17.

consideration is given to changing products to keep pace with international markets. New Zealand farmers do not always have highly fertile conditions for production but do have a very favourable climate and have proved themselves more than capable managers. Thus they can compete in foreign markets despite low returns after deducting transport costs. Horticulture has developed mainly to supply the home market as incomes have increased. Export trade in horticultural products has been mainly centred on apples and pears since 1911 and in more recent years on other fruits. Although exports of fruit still only amount to 1.5 per cent of the total export value the fruit-growing industry is now regarded by the Government as a further source of export earnings in the future. The development of air freighting services, devaluation and trade agreements with Australia have given a boost to exports of some other fruits, for example strawberries, Tamarillo *(Cyphomandra betaurea)*, Chinese gooseberries or 'Kiwi fruit' *(Actinidia chinensis)*, and Passion fruit *(Passiflora edulis)*. The size of this luxury market will clearly be determined by growth in incomes at home and overseas.

Case C. Taiwan

Taiwan is an island of 13 885 square miles and 13.5 million people. In 1967 the GNP per capita was £114. For the past decade growth rates have been high, about 10 per cent yearly. In 1952 agriculture accounted for 35.7 per cent of the net domestic product and other industries for 17.9 per cent. By 1968 agriculture's contribution had declined to 23.2 per cent and other industries expanded to 29.2 per cent. Although Taiwan is a predominantly agricultural country the Government have come to realise that agriculture is not only a source of food but also a source of investment capital and raw materials for the processing and manufacturing sectors.

The climate is sub-tropical to tropical but being mountainous it is very variable. Only one third of the land area is arable. Sugar and rice were the predominant products in 1952 comprising 81 per cent of total production. The introduction and expansion of speciality and horticultural crops has changed the agricultural economy considerably. Furthermore it has given rise to a food processing industry. This development is illustrated in table 1.10.

These new crops have certain characteristics which make them highly suitable for Taiwan's economic development programme. Crops from the temperate zone as well as tropical and sub-tropical ones grow successfully on the island and most will grow on a range of sites. The production and processing of the crops are intensive in their requirement of labour which is plentiful. Finally the fact that these crops now account for nearly 80 per cent of agricultural exports means that they are very important earners of foreign exchange currency.

Case D. Israel

Israel encompasses over 5 million acres, all owned in perpetuity by the nation, and which may not be bought or sold. Fragmentation of holdings is forbidden. Settlers or collective organisations may lease land for periods of 49 years. Most of the population of 2.8 million are rural dwellers and live in collective settlements and 12 per cent are engaged in agriculture.

The State of Israel became a reality in 1947 and early Jewish settlers were very concerned with Zionist ideals. One of these was the achievement

Table 1.10

Export value of major commodities from Taiwan
(US $1000)

	1952	1968
Rice	23 240	11 200
Sugar (Cane)	69 684	47 300
Citronella oil	2 448	2 000
Tea	5 745	11 850
Bananas	6 634	57 600
Vegetables (fresh)	151	12 000
Fruits (fresh)	771	7 000
Bamboo shoots	608	5 000
Pineapple (canned)	2 012	19 800
Mushrooms (canned)	—	32 700
Asparagus (canned)	—	28 200
Fruits (preserved)	184	10 000

Luh, C. H. (1970). 'Taiwan's success with export crops'. *World farming*, 12, 9, 2-8.

Table 1.11

Agricultural development in Israel 1949-69

	1949	1959	1967	1969
Agriculture's share of the GNP %	11.5(1950)	16.9(1958)	9.6(1966)	8.0
Irrigated area 1000 hectares	205	n.a.†	n.a.†	420
Total agricultural exports ($M)	18.1	58.1	106	110.2
Agriculture's share of total exports %	48.4(1950)	40.4(1958)	18.8(1966)	16.4
Exports of:				
(a) Citrus ($M)	18	45.9	83.7	n.a.†
(b) Fresh vegetables ($M)	—	—	1.8	n.a.†
(c) Fruit, not citrus ($M)	—	—	3.2	n.a.†
Output per worker $	14.1	25.2	43.8	n.a.†

† n.a. = not available

Ackroyd, C. R. (1971). Project Report for the Degree in Horticulture (Unpublished) University of Bath (carried out under the author's supervision).

of political stability through the establishment and maintenance of a large rural population. Strategically placed settlements were part of the defence network. Farming at first was mixed, citrus representing the sole export crop until 1954. By this time food shortages encountered as a result of large numbers of immigrant arrivals had been overcome. Thereafter agronomic changes towards greater specialisation and exportation began. Balance of payments problems heightened the need for export earnings and agriculture has been second only to cut and polished diamonds since the mid 1950s. Over the period 1949 to 1967 exports of agricultural produce increased by 600 per cent. Groundnuts and bananas joined citrus as export crops and vegetable exports began in 1954. Much of the horticultural production is dependent upon irrigation. (table 1.11).

More recently the Ministry of Agriculture have prepared an Agricultural Development Plan 1967-71 in which vegetables, fruit (other than citrus and bananas) and flowers were singled out as crops warranting the greatest expansion.

CHAPTER 2

PRICE

In a free enterprise exchange economy of the type discussed in this book the owners of resources are free to enter into production or to sell the use of the resources to others. Products and output may be sold freely. In this situation chaos might be expected to result from the lack of organisation, yet horticultural produce finds a customer either at home or overseas, factory workers are fed and clothed and the products of their labours also find a sale. Although not immediately apparent there is a system operating which some economists have referred to as the 'invisible hand'.

Each firm makes decisions about what and how to produce and sell, each consumer plans his expenditure in terms of urgent, desirable and luxury priorities. Finally the state may intervene with its own plans. Because firms are motivated towards profits and consumers towards satisfying their particular requirements there is also a system, albeit imperfect, of rewards and punishments for particular actions. It is generally known as the *price mechanism,* and results from an association of consumers and producers, having freedom of choice, in a market.

Prices are determined by the interaction of supply and demand. An understanding of this balance is of fundamental importance in the study of economics. Price itself is the money paid for particular goods or service. A grower may pay a price for an input used in production; it then becomes a cost incurred by him but represents revenue to the person he bought it from. Similarly price relationships exist when he sells the output or produce from his holding.

In the economic organisation of society as a whole, and no less so in horticulture in particular, the price mechanism has five important functions.

1. *Determination of what to produce.* Prices provide the means by which the desires of consumers are transmitted to producers. The consumer registers a preference in the market by his purchases. Having freedom of choice, he will allocate his money according to the priority of his desires. The producer advances or retards his production according to whether he finds high or low prices resulting from the sale of his produce. Thus the market is analagous to a barometer with prices as the indicator of the pressures of consumer preferences for particular commodities. If prices rise, as a result of increased interest on the part of consumers, the producer is encouraged to further production. This price increase is also an

incentive to consumers to change their purchasing from this now relatively higher priced product to other alternatives. Hence the mechanism works in two opposing ways.

2. *Organisation of production.* When firms purchase inputs for use in production they pay a price for each of them. Abundant resources may be lower priced than rarer ones. Labour, for example, in an African country may be very plentiful in relation to capital and so manpower rather than machines may be the best choice for crop harvesting. In the USA the reverse may be true in relative terms, machine harvesting assuming a greater prominence. Price in this case has determined production organisation.

A firm within an industry that receives higher prices in relation to costs than other firms is able to pay higher prices for resources. Thus they are able to divert the resources away from other firms. Some growers close to urban centres often find it difficult to compete with manufacturing industries in terms of wages. This also gives an increased advantage to efficient firms over inefficient ones, the price mechanism allowing them to allocate resources to those uses for which they have the greatest demand. In general, they will use the lowest cost resources in relation to likely returns.

3. *Distribution of production.* Producers disposing of their produce will in turn logically distribute to points where highest prices are obtained.

4. *Restriction of consumption to the stocks available.* There is always a price level which will restrict the quantity of goods and services demanded to the quantity available for disposal, so that some form of rationing process is necessary between consumers. Price performs this function. Purchases of any commodity can be increased or decreased by lowering or raising the price. This is part of the basis of indirect taxation when demand is manipulated by changing the effective price.

Considering a longer time period there will often be a price expected in the future that will make it worthwhile for the producer to postpone sales or store the product until a later date. Thus apples harvested over a few weeks may be stored until demand for apples is more favourable in relation to supply, bringing higher prices at the market.

5. *Long-term maintenance of production and development.* To ensure a continuous flow of goods and services producers must be able to see that prices provide adequate incentives to maintain their plant and equipment. The market price to the grower must satisfactorily cover the costs he incurs in the use of resources in production. If the future does not look encouraging the development of new products, or even expansion of current production, will be repressed. A common example of this is the

interest rate (price) that producers think of as their payment for the use of capital. As this capital is needed for further development or replacement investments it is not surprising that there is concern when rates are high.

None of these five factors work in isolation. All are interrelated and perform concurrently. Furthermore, they do not always maintain perfect economic order mainly because economics is concerned with human beings and their decisions. The knowledge of these decision-makers can be very variable. Communication between producers and consumers and between each other may not be good. Even after communication there are inevitable time lags in the organisation of production.

Having outlined the workings of the price mechanism it is necessary to investigate and understand the nature and origin of price as the result of the interaction of supply and demand. These are first considered separately and then in interaction. Some of the characteristics of that interaction are then discussed in relation to horticulture.

2.1 DEMAND

Demand refers to the quantity of goods or services that consumers are prepared to purchase. This is a very specific definition. It is not to be confused with need. You and I need more money, starving people need more food but can we or they afford to pay? Compare this with a statement such as 'there is a demand for good quality dessert apples at Christmas time'. Demand is always associated with a particular price of preparedness to pay, sometimes referred to as the *effective demand.*

The determinants of the demand for a product (the demand schedule) at any given time are as follows

(a) the price of the product;
(b) the consumer's income;
(c) the price of substitute and competing products;
(d) the price of complementary products;
(e) consumer tastes and fads and other variables influencing the demand for particular goods, for example, strawberry sales may be influenced by rain or temperature.

There are some new expressions in this list that are explained further below

(a) Price

When produce is expensive consumers will purchase less than when it is cheaper. This is a rational response on the part of the consumers to change in price. This relationship can be plotted graphically as in figure 2.1, a hypothetical situation. The line DD represents a demand curve or schedule and could represent a single person or even an entire population. It is important to define to whom it refers and over what time period. Taking two specific points on the curve X and Y; at X a quantity q_1 is demanded at price p_1. If the price falls to p_2 then consumers will increase their purchases to q_2. The converse is true if the price were to increase from p_2 to p_1.

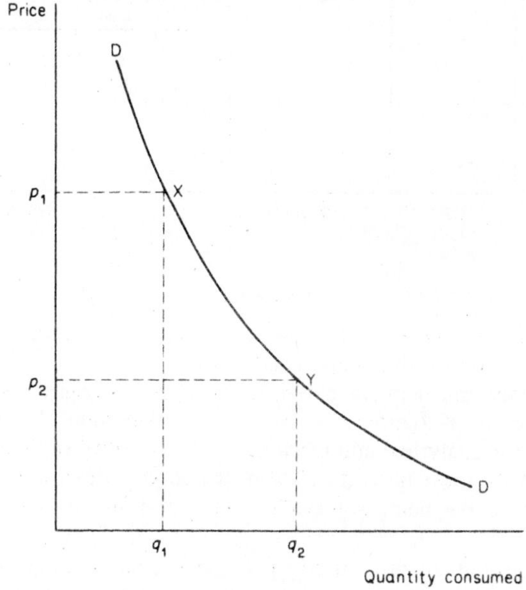

Figure 2.1

The demand curve for a particular commodity is a definable mathematical relationship and as such can be measured. In practice however, it is extremely difficult and expensive to estimate, requiring the collection of very detailed data. It is normal to assume that all other relationships remain constant (the *ceteris paribus* assumption). The rate of change in quantity consumed in relation to change in price is called the *elasticity* of the curve. When defined specifically in terms of price it is called the *price elasticity of demand.*

$$\text{Price elasticity of demand (EDp)} = \frac{\text{\% change in quantity demanded}}{\text{\% change in price}}$$

The actual elasticity for a given product depends upon all the parameters of demand and is always defined in terms of price or income and time period. If the curve had been vertical then any change in price would result in no change in quantity consumed and the curve is said to be *perfectly inelastic.* Alternatively if the curve is horizontal then consumers will buy any quantity offered at that price and the curve is then said to be *perfectly elastic.* (figure 2.2).

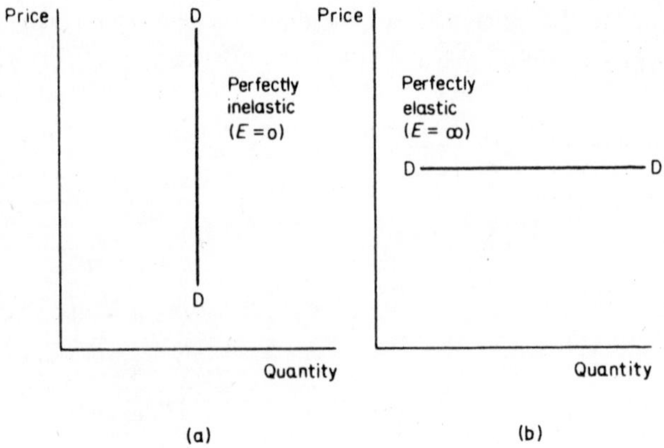

Figure 2.2

These are extremes and most commonly the curve is downward sloping; the reason for this is discussed below.

Commonly the demand curve is drawn as a straight line or even a simple curve as in figure 2.1. As an approximation this is generally satisfactory for analytical and teaching work. In reality statistical evidence suggests that for most agricultural crops the curve is downward sloping left to right, the curve being somewhat inelastic at the start but becoming more elastic.

The measurement of elasticity is not particularly complex but is very specifically defined. At every point on the curve the elasticity will differ. When the elasticity is measured over part of the curve, say XY in figure 2.1, it is called *arc* elasticity. In contrast if it is measured at a specific point say at X or Y then it is called *point* elasticity.

$$\text{EDp (arc)} = \left(\frac{q_2 - q_1}{q_2 + q_1}\right)\left(\frac{p_2 + p_1}{p_2 - p_1}\right)$$

$$\text{EDp (point)} = \frac{dq}{dp} \times \frac{p_1}{q_1} \quad \text{(at point X in figure 2.1)}$$

Figure 2.3 depicts a demand curve for a hypothetical product. To measure the arc elasticity of demand between X and Y we simply read off the relevant values from the axes of the graph and substitute in the equation above.

Thus $\left(\frac{q_2 - q_1}{q_2 + q_1}\right)\left(\frac{p_2 + p_1}{p_2 - p_1}\right) =$

$$\left(\frac{30 - 20}{30 + 20}\right)\left(\frac{4 + 7}{4 - 7}\right) = \frac{11}{-15} = -0.73$$

This index tells us that if we decrease the price by 1 per cent then the quantity consumed will only increase by 0.73 per cent. As the value is less than 1.0 it is an inelastic curve which can also be recognised from its steepness as drawn in figure 2.3.

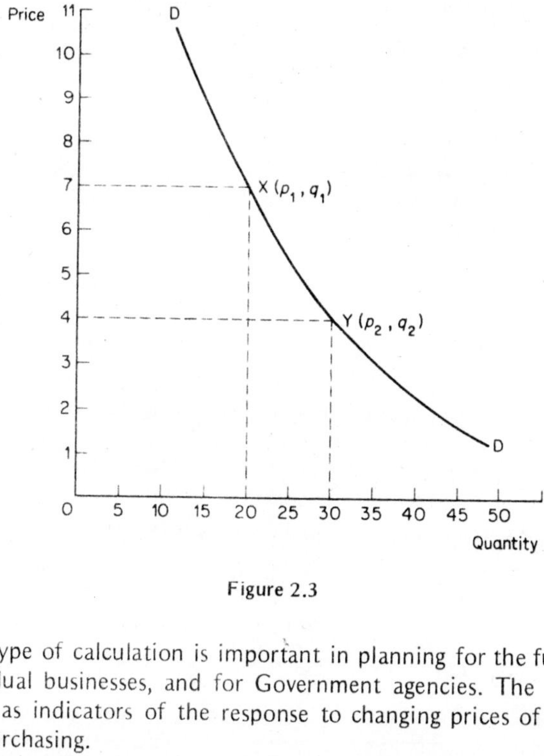

Figure 2.3

This type of calculation is important in planning for the future, both for individual businesses, and for Government agencies. The indices are important as indicators of the response to changing prices of consumers who are purchasing.

Table 2.1 presents a range of agricultural products in the UK arranged in order of magnitude of price elasticity of demand. These are aggregate figures calculated as the mean of a year's retail trade. The quantity demanded does of course change according to seasonal effects during the year. It will be further noted that there is a wide difference in the response of purchasers to price changes between commodities. This depends upon the importance and priority that is placed on the product by consumers. Potatoes are a very staple part of the diet of UK residents and for a change in price very little change in consumption will result. More luxury goods however show a greater response. The difference in elasticity between milk and cream should be noted in this respect and of course the fact that the horticultural products are predominantly in the lower half of the table, having a relatively greater response to price than agricultural products generally.

Table 2.1

Price elasticity demand for selected products in United Kingdom 1963-64
(At retail prices)

Potatoes	−0.07
Margarine	−0.23
Butter	−0.28
Milk	−0.31
Pork and lamb	−0.41
Apples	−0.56
Fresh salads	−0.88
Quick frozen peas	−1.34
Cream	−1.35
Brussels sprouts (fresh)	−1.44
Pears	−1.71
Cauliflower	−2.45

HMSO (1965). 'Household Food Consumption and Expenditure'.

In table 2.2 actual purchases in the UK are compared in the same time period as the elasticity indices of table 2.1. Using the average of the years 1958 - 64 as a base, year to year fluctuations even out and percentage changes from month to month have been calculated. The marked summer consumption of salad crops, in contrast to the winter consumption of carrots, and the more regular eating of apples is evident.

Table 2.2

The pattern of seasonal purchases of salads, apples and carrots in the UK. Actual purchases per head
(Average 1958-64 = 100)

	Jan	*Feb*	*Mar*	*Apr*	*May*	*June*	*Jul*	*Aug*	*Sept*	*Oct*	*Nov*	*Dec*
Leafy salads	30	41	80	161	245	288	257	191	148	78	48	33
Fresh carrots	144	128	124	103	75	62	61	73	96	118	137	133
Apples	95	110	105	96	99	88	77	86	109	120	117	109

HMSO (1966). 'Examination of the Horticultural Industry 1965'.

(b) Incomes

The consumer is said to gain satisfaction or utility from a product, utility being the capacity of goods or services to satisfy a desire. Unfortunately the word utility has other connotations of usefulness and desirability so that confusion results from its use. In modern times therefore it has tended to be superseded by the expression 'preference'. Students new to the social sciences but familiar with measurement in other sciences will no doubt recognise at this point the very real problem of measuring concepts like preference and satisfaction.

All consumers have a limited income at some level and will have a scale of preferences for the purchase of any product. There is an upper limit to the amount of a product that can be consumed, with food the biological limitations in capacity of the human stomach. As more is consumed there is a progressively diminishing satisfaction obtained from consumption. A thirsty man may derive great satisfaction from drinking a glass of water. A second glass may be very acceptable but he will derive less satisfaction from it than he did from the first. Thus the satisfaction from an additional quantity consumed depends upon how much we have to start with. This is the so-called *law of diminishing marginal utility (or satisfaction).* It says that any increase in a person's rate of consumption of a commodity reduces the intensity with which further increments are demanded.

The 'marginal' concept in economics has now been introduced. We return to it again in chapter 4. In this case we are concerned with the extra (marginal) benefit received from an extra (marginal) unit consumed. Total utility or satisfaction to the consumer rises to a maximum point and then declines in the manner of a response curve. The marginal utility curve in contrast declines continually. These curves are represented in figure 2.4.

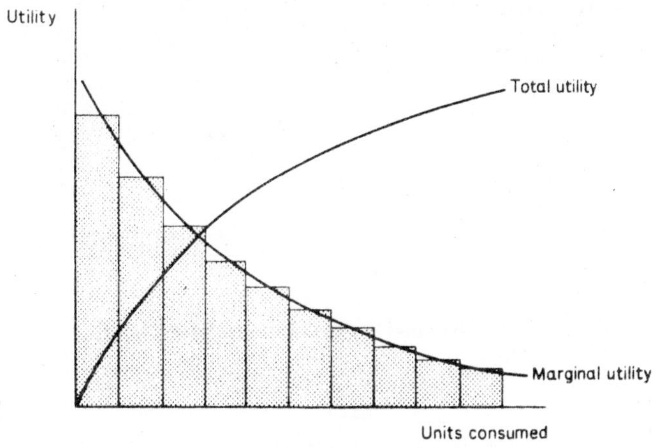

Figure 2.4

If the consumer allocates his income to total utility it will not be worthwhile for him to transfer to any other goods – the marginal utilities will be equal. This is an old but important concept in economics and explains further the general downward slope of the demand curve.

As incomes increase consumption generally increases but the proportion of the components of consumption may change at different rates. The consumer may change the priorities of his spending as incomes increase. As we have already seen for price changes it is possible to measure in quantitative terms the amount of change of purchasing that will result from a particular change in income. It is this time called *income elasticity* of demand and is defined thus

$$\text{Income elasticity of demand } (ED_1) = \frac{\text{\% change in quantity consumed}}{\text{\% change in consumer income}}$$

Tables 2.3 and 2.4 present, as examples, income elasticity indices for the UK and some international comparisons. Table 2.3 shows clearly change in elasticity over a five year period for several horticultural products. Table 2.4 considers broader groupings of commodities in

Table 2.3

Income elasticities of demand for selected products UK 1962-67. (Actual expenditure)

	1962	1967
Potatoes (all)	−0.06	−0.04
Cabbage	0.11	0.21
Brussels Sprouts (fresh)	0.42	0.34
Cauliflower	0.60	0.38
Leafy salads	0.94	0.56
Peas (fresh)	0.64	0.41
Peas (quick frozen)	1.34	0.91
Cucumber	n.a.†	0.68
Oranges	0.78	0.55
Apples	0.84	0.58
Pears	0.95	0.65
Soft fruits	1.18	1.30
Rhubarb	n.a.†	0.39
Tomatoes	0.45	0.40

† n.a. = not available

HMSO 'Household Food Consumption and Expenditure, 1962 and 1967'.

Table 2.4

Income elasticities of demand in different countries

Region	*Fruit and vegetables*	*Meat*	*Cereals*
USA	0.25	0.35	−0.5
Australia	0.4	0.1	−0.5
EEC	0.6	0.7	−0.3
UK	0.5	0.4	−0.4
Near East and Africa	0.7	1.3	0.1
Latin America	0.55	0.75	0.14

FAO 'Agricultural Commodities Projection for 1970'. *Commodity Review* 1962, special supplement.

countries at different stages of development. The higher response of horticultural products and meat to higher incomes is evident.

It was Engel who put forward the theory now known as Engel's law that as incomes increase the percentage of them spent on food decreases and a larger proportion is spent on luxuries. The importance of increasing incomes to horticultural expansion has already been stressed in chapter 1. Within specific countries the division between purchases as incomes increase is important. In the so-called developed countries expenditure on food has fallen to a relatively small proportion of total income and recreation activities and non-food luxuries have taken on greater prominence. Only 20 per cent of UK incomes are now spent on food. Another very important trend as incomes increase is the demand for more services associated with food. In the case of horticulture these may take the form of ready-prepared meals requiring only heating or simply presentation of the produce ready cleaned or peeled for cooking. These are the so-called 'convenience foods' which are increasingly being purchased, although the actual quantity of food consumed is not increased. Table 2.5 illustrates the increase in expenditure on convenience foods in the UK in recent years.

Table 2.5

Indices of expenditure on convenience foods in UK 1962-67
(1963 = 100)†

	1962	1963	1964	1965	1966	1967
Canned peas	97.3	100	94.3	91.2	91.6	94.6
Canned beans	91.7	100	103.8	108.3	111.3	121.5
Canned and bottled tomatoes	92.7	100	138.2	145.5	149.1	157.1
Quick frozen peas and beans	80.2	100	85.5	87.2	111.9	108.7
All quick frozen convenience foods	86.7	100	100.0	110.3	139.1	139.7
Total all convenience foods	96.3	100	104.5	109.4	117.9	124.3

† Severe weather in late winter and spring led to a high consumption of convenience foods in this year.

HMSO (1969). 'Household Food Consumption and Expenditure, 1967'.

(c) Substitutes and complementary products

The consumer when purchasing chooses between the range of products available so that he spends his limited income to his greatest satisfaction. Certain priorities will always take precedence and the pattern of desires will further be influenced by his upbringing, social factors, even the degree of salesmanship encountered. With these variables involved it is clear that we cannot consider what the consumer ought to buy but only what he actually bought in given time periods and situations.

For any two products it is possible to establish a relationship between them as to the level of satisfaction achieved from different combinations of purchases. When plotted graphically they are called *indifference curves* – they are curves joining points between which a consumer is indifferent because they have the same level of satisfaction for him. Points A and B in figure 2.5 represent points of indifference, they are both on the indifference curves of a consumer to two products, apples and oranges. The consumer would be equally satisfied by X_1 units of apples and Y_1 units of oranges as he would by X_2 units of apples and Y_2 units of oranges. We say that the products are *substitutes* for each other because the consumer will give up one for the other according to the relationship defined by the indifference curve.

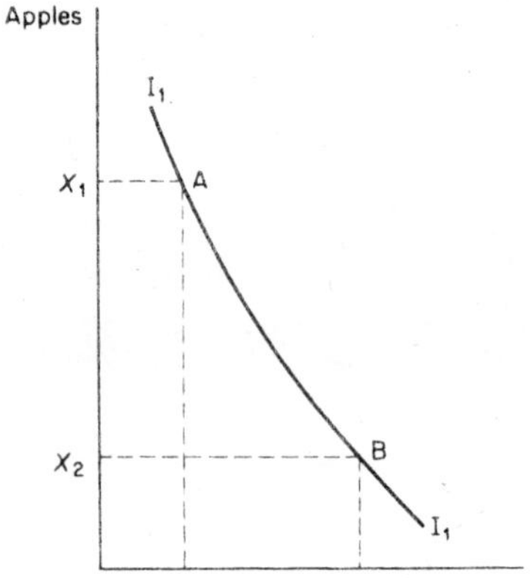

Figure 2.5

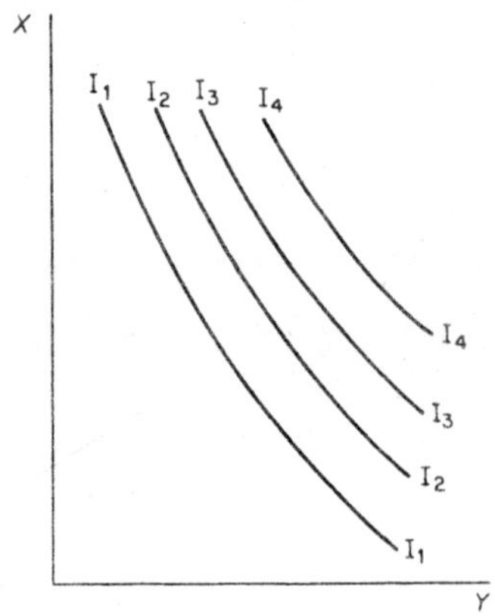

Figure 2.6

Multiple indifference curves may be drawn as in figure 2.6. I_2 representing a greater total satisfaction than I_1. There is an obvious analogy here to contour maps but we cannot measure the distance between curves and usefully analyse gradients between them. Further characteristics of indifference curves are that they cannot cross and are negatively sloped, convex to the origin. All other representations of indifference are in conflict with reality.

So far we have considered qualitative relationships only with respect to indifference curves. To evaluate how the rational consumer will allocate his limited income between the products it is necessary to introduce price data. Supposing in the situation depicted in figure 2.6, X costs £1 and Y £2 per unit and the consumer has £10 to spend. Y costs twice as much as

X and so the maximum quantity of each that can be purchased alone is 10 and 5 respectively. This is represented graphically in figure 2.7 by the line PP called the price line or budget line. Amongst the family of indifference curves there will be one which just touches the budget line, I_2 in this case. At the point of tangency satisfaction is maximal, that is when $\Delta X/\Delta Y = PX/PY$. At no other point is satisfaction maximal. It is the only point for a given budget, at which the consumer can get on to the indifference curve and indeed if he purchases anything below the price line, he is not using all his income. It is important to realise that we have considered a static situation; in the real world all the factors involved are changing continually and choice is often between more than two products and it is a dynamic process.

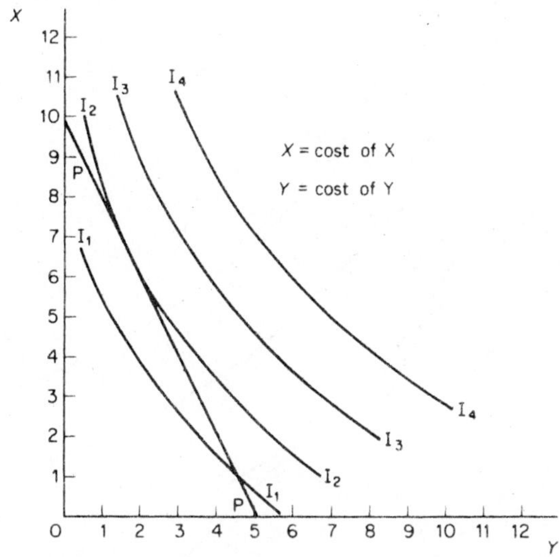

Figure 2.7

Consumer income and product price can be expected to change frequently. If the consumer's income increases then he has more money to spend and his budget line moves upwards from AB to CD in figure 2.8. He can now afford more of both goods. The new budget line will be tangential to a new indifference curve and a new point of maximum satisfaction identified at *v* above *u*. This representation assumes that all other factors remain static.

If the price of one product changes in relation to the other then the slope of the budget line will change. This might be represented by the line CE in figure 2.8. Y has become less costly in relation to X. A new point of tangency to indifference curve I is established at *w*. These effects are called *income* and *price effect* respectively.

If an increase in demand for a product leads to an increase in demand for another we say that there is a complementary relationship between them. This is in contrast to the substitution situation that we have just been considering where there is competition between products for the consumer's income. Good horticultural examples of complementary products are to be found in the purchasing of vegetable salads. When the housewife buys these she is concerned with providing a balanced meal and rarely purchases them singly. She may buy tomato, cucumber, radish, lettuce, salad onion, beetroot and pepper at the same time. Strawberries and cream are another example, often quoted in textbooks, of complementary products.

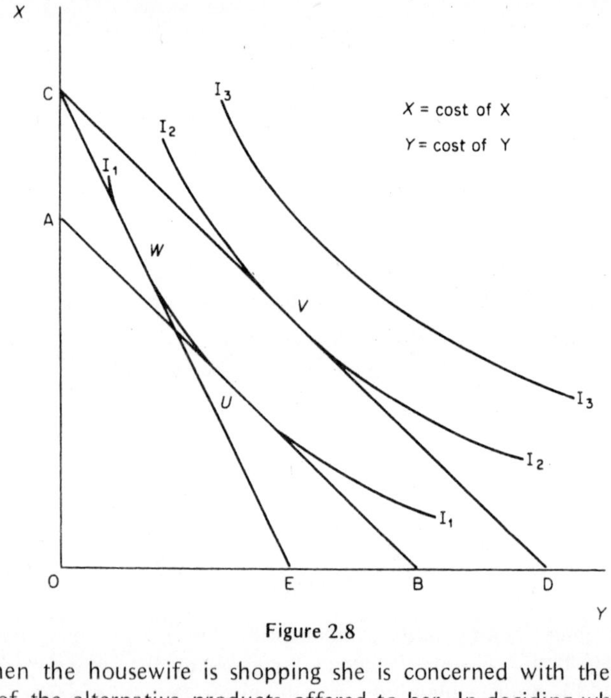

Figure 2.8

When the housewife is shopping she is concerned with the relative prices of the alternative products offered to her. In deciding whether to buy brussels sprouts or cauliflower she will consider amongst other things the relative prices of them. If the price of cauliflowers should fall, and the price of brussels sprouts remain unchanged, she will be more likely to purchase cauliflower. A fall in the relative price of a commodity causes a transference of purchase (substitution) between commodities and is called cross elasticity of demand when its rate is measured. The closer the substitutes the greater is the likelihood of transference by the consumer. Brussels sprouts and cauliflower, for example, are closer substitutes than apples and potatoes.

We have now considered the main factors influencing the demand for our products. Considerable space has been devoted to this because of its importance, which will become more apparent in the next chapter when we consider marketing. It is the consumer who ultimately decides whether we sell our products. The old adage that 'the customer is always right' is still as true as ever.

2.2 SUPPLY

Total supply of produce is the supply from all producers arriving at the market. Normally it is expected that the supply increases as price increases (figure 2.9) but this is not always true in reality. At price p_1 a quantity of product q_1 is sent to market. If the price is raised to p_2 then the producer (S) will increase supply to q_2.

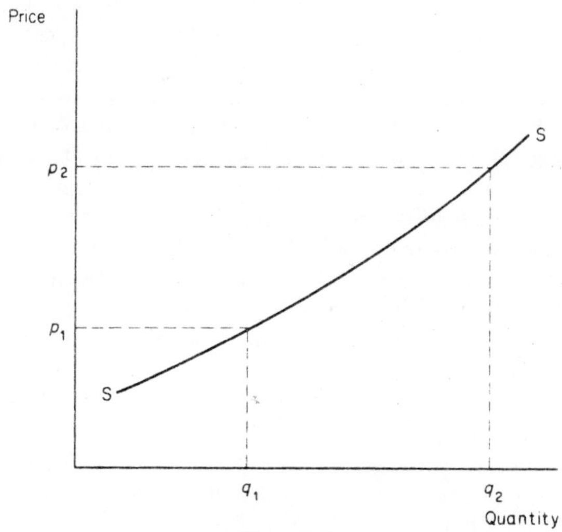

Figure 2.9

As for demand curves, it is possible to measure the slope of the curve and this is expressed as the *elasticity of supply* in terms of price.

$$\text{Price elasticity of supply (ESp)} = \frac{\text{\% change in quantity supplied}}{\text{\% change in price of product}}$$

If the elasticity index is greater than 1.0 then the curve is elastic, if less than 1.0 inelastic. Normally it is expected that the elasticity will be positive, the line sloping upwards left to right.

For a whole agricultural sector of an economy with a vast number of holdings the elasticity for many farm products is very high. Even small increases in price will give rise to a large increase in production, and often in the number of producers. Horticulture has a particular problem in this respect arising from the length of the period required to produce crops which in turn are often determined by the climate of the particular country. For annual crops area or output can be increased or reduced each year. Even so an increase in price within a growing season cannot always bring forth sufficient quantity to meet needs. This is because the total supply in any one year cannot exceed the total yield of all producers together with any carry-over from the previous year. Yields are dependent upon the amount actually sown or planted and weather and management thereafter. Imports may compete in the home market and fill any gap in supply left by home producers.

A grower who has established plantation crops as bushes with a life of say 10-15 years or tree crops with a life of 40-50 years is not so easily able to adjust to changing markets. Even if he scraps his investment before the end of its natural life he is likely to have kept it long enough to test its viability. A glasshouse grower, after erecting his houses, is unlikely to cease this type of production for some years. He does have greater flexibility than a grower of unprotected crops in that he may be growing as many as three crops per annum and can change these. For such longer term crops or ventures the grower's estimation of the future may well impede his rational supply in the shorter term. Once he is committed he tends to increase production in response to a price fall rather than decrease it. He tries to sell more to maintain his income. The supply of cherries in the UK is of particular interest in this respect. Prices have remained fairly stable and have even increased slightly but the acreage and production of cherries is falling rapidly for reasons of income uncertainty due to variable yields and cultural problems. There is considerable uncertainty too about the future for cherries leading to little reinvestment in the crop. This type of relationship where supply does not increase with price is called *perverse*. Agriculture has many more examples of this phenomenon usually associated with particular characteristics of farming practice.

In a general depression when prices of all agricultural products fall growers have the choice of going bankrupt, producing more output and accepting a lower net profit or leaving the industry altogether. Generally growers are self-employed and cannot stand themselves or their labour force off as is done in the manufacturing sectors in such a situation. Countries that depend upon export trade may suffer severely when this happens.

Perishability is a characteristic of horticultural produce. Strawberries and lettuce for example are saleable for only a matter of hours in some instances, at best a few days, so that if the market is over-supplied they become unsaleable. In the mid-season of outdoor cropping the supply of strawberries will be almost totally inelastic because the crop is ready to be sent to market. If this coincides with particularly good weather, demand may be stimulated but it accentuates the problems of over-supply and much of the crop may not reach a consumer.

2.3 INTERACTION OF DEMAND AND SUPPLY

At the market demand and supply come together and a price is determined. The concept of the market is discussed in the next chapter. In figure 2.10 supply and demand curves have been superimposed. At the point where they cross (O) they are equal and are in equilibrium. Supply and demand are constantly changing so that the point of equilibrium also changes. As it has been drawn the mechanism has been arrested at a moment in time.

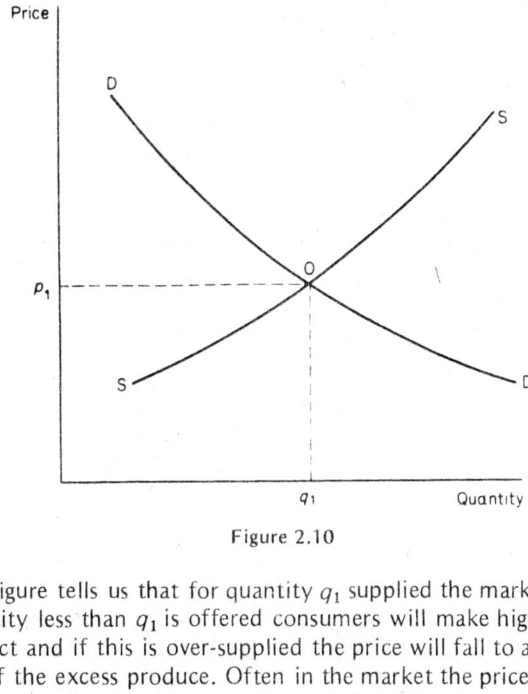

Figure 2.10

The figure tells us that for quantity q_1 supplied the market price is p_1. If a quantity less than q_1 is offered consumers will make higher offers for the product and if this is over-supplied the price will fall to accommodate the sale of the excess produce. Often in the market the price will oscillate around the equilibrium point particularly if dealings have begun at some other point than the true equilibrium.

Varying demand and supply

Let us consider the price of cabbage. In figure 2.11(a) an equilibrium exists at price, p_1 for a quantity q_1 supplied. Supposing that consumers, growing tired of cabbage, transfer their purchases to other vegetables as they become cheaper in summer-time, the demand for cabbage falls and the curve shifts bodily to the left to D_1D_1. The quantity consumers are now prepared to buy is q_2 and the price at the new equilibrium point is p_2. The supply curve has remained unchanged; this is quite a realistic assumption because growers will have planted the crop some months

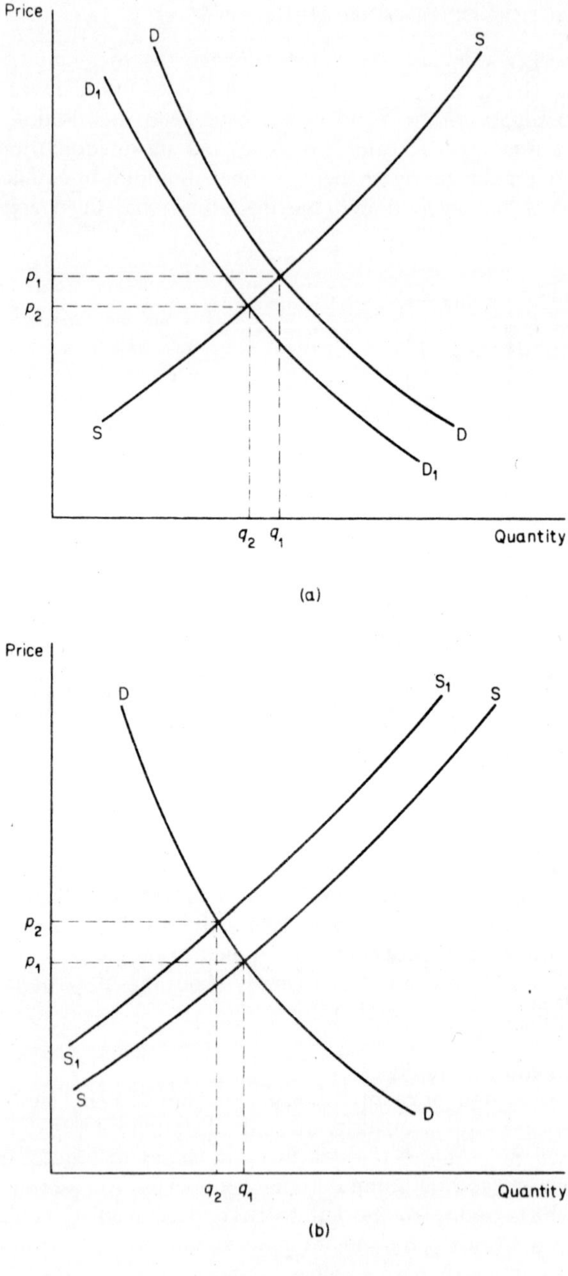

Figure 2.11(a) and (b)

earlier. As time goes on the growers may decide to keep supplying, or when the price falls to an unacceptable level simply 'plough-in' the cabbage. If a sudden period of freezing weather should make transport to market and even harvesting the cabbage impossible, the supply at the market diminishes. This is the situation in figure 2.11(b). Demand for cabbages remains steady but supply has decreased from SS to S_1S_1. The supply curve in this case has moved bodily to the left and a new equilibrium is established. The price has risen from p_1 to p_2. These movements go on continuously with changes in supply and demand. The reader may like to consider for himself, at this stage, the effect of (i) increasing demand in situation 2.11(a) and (ii) increasing supply in situation 2.11(b).

Cyclic prices – 'Cobwebs'

A common real world result of the interaction of demand and supply is a cyclic pattern of production. Horticulture is characterised by this type of response, perhaps annual outdoor crops being the most susceptible.

Let us consider a carrot grower who in a given season receives a high price relative to other years for his carrots. The following season he sows more carrots as a response to the higher price. Other carrot growers receiving the same stimulus also increase their sowings. The season is a particularly good one for carrot growing and yields are high, so the aggregate production of all growers is considerably in excess of the previous year and prices fall to a lower level. Growers are dissatisfied with the price and transfer some of their acreage in the following season to other crops, reducing the total crop marketed. The price improves and the cycle starts again. This effect is particularly common when crops have similarities in their cultural requirements and machinery. Outdoor vegetables are a good example and cycles tend to be short. For soft fruits the cycles are longer because of the longer investment period involved. Figure 2.12(a) and (b) demonstrate graphically typical cyclic behaviour.

A grower at equilibrium A in figure 2.12(a) sees a higher price for his product in the future and increases his supply to B. At this quantity however, demand is only C and the grower gets a lower price than expected and reduces his production to D to match demand. Other growers having done the same, the market is now under supplied and so price is reflected at E encouraging expansion to F. In this example the supply curve has been deliberately drawn more elastic than the demand curve and it gives rise to a *diverging 'cobweb'*. The name cobweb is derived from the appearance of the graphical diagram. In figure 2.12(b) the demand curve is more elastic than the supply curve and for the same series of events the diagram now becomes a *converging 'cobweb'*. These cobwebs are generally arrested by a new combination of supply/demand schedules being derived but then start all over again. Often fluctuation is continuous but contained within definite quantity ranges.

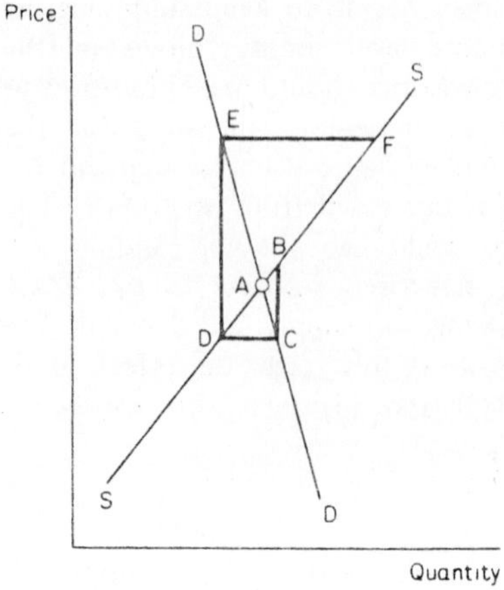

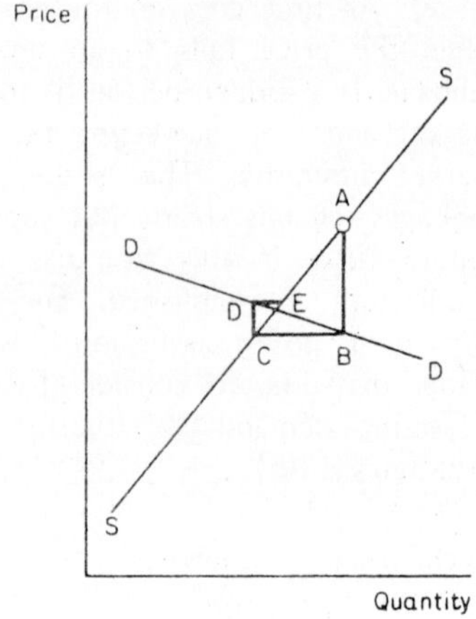

Figure 2.12(a) and (b)

Table 2.6 presents data for outdoor beetroot *(Beta vulgaris)* production in England and Wales. Distinct cyclic patterns show in the total production and prices recorded. The pattern of production can be further analysed by considering the components of yield and acreage grown. The student will find it worthwhile to try to explain each year's data in the light of the situation in the preceding year.

Table 2.6

The production and marketing of beetroot in England and Wales, 1959-60 to 1966-67

Year	*Acreage (1000)*	*Yield (tons per acre)*	*Production* 1000 *tons*	*Av. annual price per hundredweight at wholesale* £
1959-60	7.5	9.7	72	1.22
1960-61	9.9	12.7	126	0.91
1961-62	7.1	11.1	79	1.17
1962-63	8.0	10.9	87	1.29
1963-64	8.5	12.1	103	1.06
1964-65	7.9	13.3	105	0.98
1965-66	6.4	13.1	84	1.24
1966-67	7.2	13.3	96	1.28

Adapted from HMSO, 'MAFF. Agricultural Statistics.'

Further illustrations of these and other cycles on an acreage basis are presented in figure 2.13. All the crops except most blackcurrants are sold in conditions of near perfect competition where there are many buyers and sellers, none of whom can individually influence the general price level. Blackcurrants are grown predominantly for processing under contract and the general upward trend and then downward since 1963-64 reflects the incidence of contract agreements. The relatively longer cycles of the soft-fruits in relation to the annual vegetable crops is obvious.

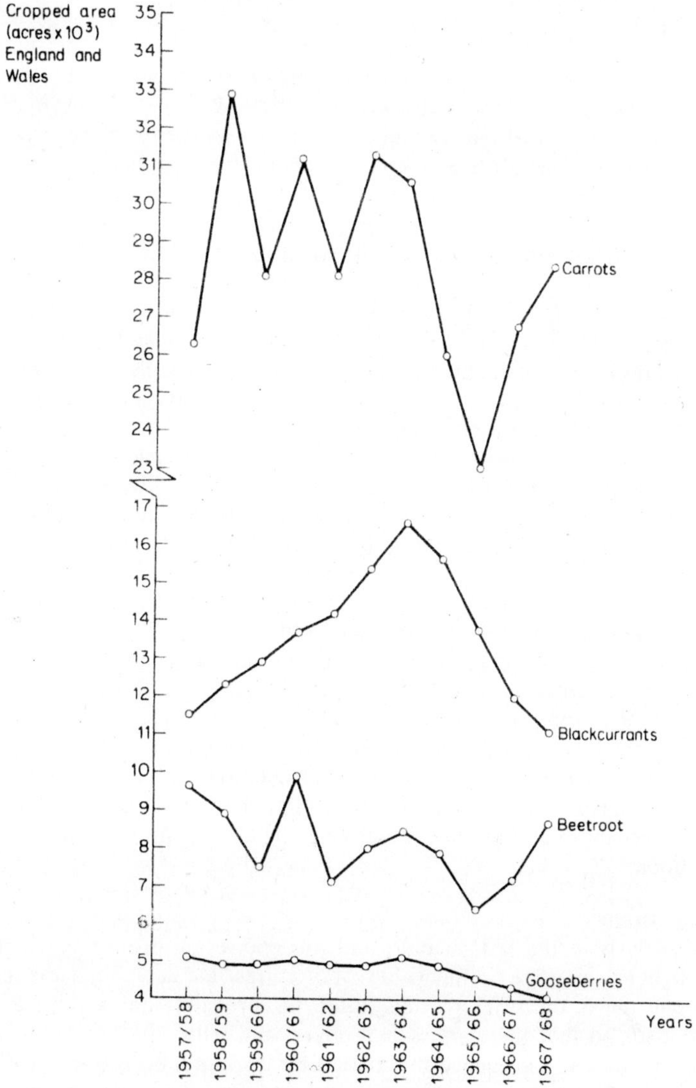

Figure 2.13

CHAPTER 3

MARKETING

To an economist a market is any coming together of demand and supply. This may not necessarily be a place although many prices are still arrived at in market places, fairs and other exchanges. We talk of 'the market' where the prices of labour, capital and land that we use for production are determined. Refrigerators and cars for example are not sold in a market because in both instances the manufacturer may recommend the price by adding to the cost of production a margin to allow for sales and distribution costs and profits. This does not mean that he can charge what he likes for he is still in competition with other manufacturers to satisfy the consumers' desires. By contrast most horticultural produce is sold in what is called a free market exchange situation, supply and demand being dominant in determining prices. There are characteristically a large number of buyers and sellers none of whom can individually exercise much influence on the general price levels.

The process of *marketing* is that of transferring goods and services from producer to consumer at profit. This is quite a different definition from that above. It is concerned with the process of adding the maximum value to the product at minimum cost. The marketing organiser is concerned with such activities as transport, selling, advertising, packaging, market research, planning and accounting; in short with the processes which enable him to maximise the value added.

Marketing is not an operation to consider as an after-thought to producing a crop. It is, or should be, the concern of all commercial growers. Marketing is a central management function and cropping should only proceed where there is an awareness of when and how the product is to be marketed. Effective planning of production can only result from a clear market objective. It is for this reason that we have considered the price mechanism and marketing before going on to consider production in this book.

Competition

Our analysis in the last chapter, and this one so far, assumes a situation throughout of pure competition and a free market. Fundamentally, competition is the offer of substitutes. In our supply/demand model we have assumed that there are several buyers and sellers, no firm being large enough to unduly influence the market by its output decisions. It follows that there is no incentive to the seller to restrict his supply to influence the market or to the buyers to restrict purchases and thus lower prices. This is *pure competition.* Any other form is called *imperfect competition.*

When there is only one seller of a product with no close substitutes then price manipulation is clearly possible and a *monopoly* is said to exist. The word monopoly has taken on an odious association in the minds of many. This stems mainly from a belief that this power is always abused. Nationalised industries selling coal, electricity, gas or rail services are some examples of monopoly sellers. It is important however, to realise that a firm can only be a monopolist for a *single* product. In the UK the National Coal Board has a virtual monopoly on large-scale coal working but it does not have powers to control all fuels so it is still in competition with other fuel producers. If sufficient statutory powers are given to agricultural marketing boards they can operate as monopolists.

Most countries legislate against the emergence of monopolistic firms. In the UK the Monopolies Commission sits to consider impending monopoly situations and there is also a Restrictive Practices Court. The USA has had Anti-trust laws for over 75 years for the same purpose.

If a few firms are involved in selling and although still in competition with each other are able individually to influence the market by their decisions, then we say that an *oligopoly* exists. Sales of fertilisers, crop protection materials and fuel oils are examples of products commonly bought by growers from this type of firm. If there is only one, or a few buyers of a product, then we say that either a *monopsony* or an *oligopsony* exists. The banana crop in the Windward Islands for example is handled by a single firm and 80 per cent of the Jamaican crop by another. Both supply the UK predominantly. Firms in an oligopolistic position are usually able to identify the percentage share of the total market that they possess; this is a rare situation for the individual horticultural producer.

For most agricultural and horticultural products virtually pure competition exists. Consumers' competition is inevitably free, the average housewife competing with millions of other housewives. It is inconceivable that her purchasing may influence the overall price. On the supply side there are characteristically many growers and for the more staple lines such as potatoes, brassicas and apples it is unlikely that a decision by one of them will unduly affect market prices. A large scale specialist producer of a more luxury product might however influence the market to some degree according to the size of his production and the market.

Marketing creates utility

That marketing is concerned with maximising the value added we have already established. We can go farther than this and say that marketing also creates value. A field of currants, apples or oranges has no value until the fruit is exchanged for cash. It is marketing that brings this about. In maximising the value added marketing also achieves other useful services to the consumer and society in general. We call these utilities. Marketing creates four of them.

Firstly, marketing creates utility with respect to *place*. Consumers are generally geographically separated from production; a field of strawberries is of little value to consumers 50 miles away in a city. Marketing transfers the strawberries to a point where they are accessible to consumers. Secondly, marketing creates utility with respect to *time*. In a seasonal climate, consumers may wish to eat fruit and vegetables all the year round. Clearly provision has to be made to store or preserve the product for the off-season. Imports might fill the gap but storage may still be necessary during transportation. Western European consumers are able to purchase apples all the year round by storing their own crops and by the importation of apples from the southern hemisphere transported under carefully controlled conditions. Thirdly there is *form*. We have already seen how, as incomes increase, consumers are prepared to buy more services with their food. The busy, often working housewife of the developed countries is more interested in the purchase of ready washed vegetables, frozen ready to cook meals, prepackaged food or graded produce. In all these instances the form of the product has been changed in the marketing process to facilitate the satisfaction of the consumer. The fourth utility is that of *ownership*. Title deed to a product may pass through several hands en-route to the consumer, for instance those of wholesalers and retailers. Marketing has the distributive function that enables ownership to be achieved. We could summarise this section by saying that nothing has value unless it is available in the right form, at the appropriate place and time.

Until a society develops to the point where its population begins to specialise in production there is no basis for marketing. The subsistence farmer may be only concerned with feeding and maintaining himself and his family; he is not concerned with marketing. As the markets evolved they have always been and still are the point of contact of buyer-seller or producer-consumer. Until recent years economic theory and marketeers only considered the situation of pure competition and a market for a product at a given level of demand. This attitude has changed somewhat with a greater tendency towards imperfect competition. It is now common in developed countries to try to influence or manipulate demand. Producers of non-agricultural goods may well create a demand for their product. A toy manufacturer may invent and produce a new toy and then spend vast sums of money convincing children and parents that they wish to purchase it. They do this by the use of publicity, advertising, demonstration and other tools of the marketeer. Even growers, despite the fact that much of their produce is sold in nearly pure market circumstances, are today being forced into at least trying to influence demand.

Before considering the different channels of distribution and institutions concerned in horticultural marketing it is worth pondering on what it is that we require from an efficient marketing system in relation to the nature of our products.

What constitutes an efficient marketing system?

Eight requirements are listed below but this section is not an exhaustive statement of how to run a market system. It is more of an attempt to provide the reader with a framework so that he can assess particular systems encountered and critically read the remainder of this chapter.

1. Consumer demand should be reflected quickly and effectively through each stage of distribution to the producer.
2. Any major changes in the costs, availability of services, and supply of particular commodities should be reflected quickly and effectively to the consumers.
3. With respect to the kind and quality of service required by consumers marketing and distribution services should be supplied at minimal cost per unit handled.
4. Uncertainty and even misallocation of supplies as a direct result of the instability of producers' and distributors' incomes should be minimised.
5. To maintain progress the marketing system should be flexible so that innovation, experiment and development are encouraged and rewarded.
6. Bargaining and price determination should be equitable to both buyers and sellers.
7. Adequate, factual, market information should be collected and made available to allow all concerned in marketing from producer to consumer to perform their function efficiently.
8. Facilities should be provided so that at each stage the quality and value of produce can be assessed.

Characteristics of horticultural products

(a) Like all food products they are perishable, some having an extremely short useful life once harvested. Lettuce and strawberries may only last for a matter of hours at high temperature. Oranges and apples by contrast have a rather longer shelf life but still need to be protected from desiccation and microbial attack.

(b) Horticultural produce is bulky in relation to its value. This is important in transportation. To protect flowers for example from physical damage, a much larger container is required than the contents fully occupy. Horticultural products show wide differences in this respect. Compare two identical lorries, one loaded with apples and the other with cabbages: the value of the former is very much larger than the latter.

(c) Often the production areas are geographically separate from the population centres where the main consumption occurs. In the early stages of development horticulture usually emerges around the periphery of these centres but as transport networks grow it is common for horticultural production to become located at the most favourable sites which seldom coincide with the manufacturing industrial centres. In the UK most of the population lives in a band of conurbations running south east to north west across the country. The main horticultural areas are located away from this band. Also the main ports of entry for imports are situated around the coastline, making further transportation on land inevitable.

(d) The supply of produce from day to day, and place to place, is characteristically variable because of the influence of seasons and local weather which often make planning of supply difficult.

(e) Unlike the manufacturer who installs a machine and produces a standard product thereafter, the grower uses biological material. Apples do not grow in a single size, colour and quality, even though much research is directed towards this goal. Grading has to be employed as a means of attaining a standardised product.

(f) Production is commonly in the hands of many uncoordinated producers. To achieve large scale coordinated organisation they may need to be united towards some specific marketing aim. For example the New Zealand Apple and Pear Marketing Board acts as the sole exporting agent for New Zealand deciduous fruit.

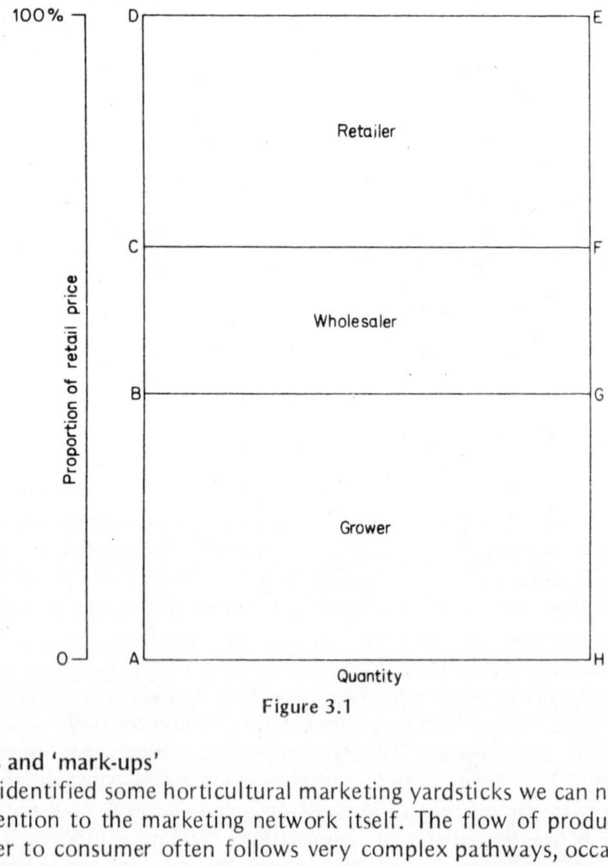

Figure 3.1

Margins and 'mark-ups'

Having identified some horticultural marketing yardsticks we can now turn our attention to the marketing network itself. The flow of produce from producer to consumer often follows very complex pathways, occasionally very simple ones. Customarily the produce passes through several hands before final consumption. Each of these 'middlemen' will require payment for the marketing service that he bestows. All stages must take their payment from the final value of the produce – the price that the consumer pays for the product immediately before consumption. Put another way, the value at retail represents the result of the additional values added at each stage of distribution. The part of this total that accrues to each organisation involved in the marketing chain is called the margin. Diagramatically figure 3.1 represents a situation where produce is handled by a wholesaler and retailer between grower and consumer. The quantity of produce AH remains constant. The area ADEH represents the total price paid by the consumers. The grower receives ABGH. The two other blocks represent the margins taken or received by the retailer and wholesaler respectively. If a wholesaler sells on commission, he does not take title deed to the produce but merely distributes it, taking a percentage commission as payment; he is called a primary wholesaler. If however he actually buys the produce and sells it again, adding on a percentage to cover handling and profit, then he is a secondary wholesaler and he 'marks-up' the produce. The retailer almost always buys and adds on a 'mark-up'.

It is important to realise competition exists throughout the chain and retailers compete with each other in the High Street and wholesalers in the market. This provides a natural brake and regulator, not allowing margins to rise or to stay excessively high.

In figure 3.1 the grower appears to receive only about 40 per cent of the total. The actual proportion received by growers has been estimated by several research organisations but it is an intangible figure to evaluate accurately at any moment of time. In the USA farmers and growers probably get 35-40 per cent for all agricultural produce and in the UK well below 50 per cent.

Integration

With over 50 per cent of value passing to the distributing organisations it is not surprising that attempts are made to streamline and eliminate some of the stages. The obvious advantage is the possibility of receiving a larger share. The chain may be shortened for example by eliminating the wholesaling function in figure 3.1 and a large retailer buying direct from a grower or grower organisation. Then the total margin is shared only by the retailer and grower but not necessarily equally. This process is called *vertical integration* but some workers prefer to use the word coordination rather than integration, true integration resulting when firms merge, amalgamate or are taken over. In vertical coordination each firm retains control of its own activities. If the number of organisations in the marketing chain increases it is called vertical disintegration.

Integration occurs laterally too and is then called horizontal. Examples might include growers forming a cooperative organisation to market their produce or a group of retailers agreeing to buy collectively from a wholesale market. In both instances the effective number of individuals involved in marketing is reduced.

For the remainder of the chapter we consider the important characteristics and some problems associated with the main methods of marketing horticultural produce. Firstly we consider the wholesale markets, including auctions; secondly, grower organisations and lastly aspects of the retail trade.

3.1 THE MARKETS

(a) Wholesale markets

Market sites usually arise, or are established, in conurbations or in convenient positions for potential customers in the country or in ports. This is certainly true in the UK. There is a clear link to economic development and in particular the emergence of horticulture as an industry. Until urban centres are large enough and the population sufficiently wealthy the conditions for the emergence of a wholesale market are not met. In the UK about three quarters of horticultural produce still passes through wholesale markets but rapid changes are occurring. As the effects of integration between producers and retailers are felt the markets decline in overall importance in terms of the volume of produce handled but they still retain the crucial roles of determining prices and standards for their own and other marketing channels. The overall dominance of this type of market in horticulture means that it warrants lengthier treatment here.

Market organisation

Specific organisation varies from country to country. Generally however, produce is consigned to the market by growers or organisations on their behalf, usually by prior arrangement with the wholesaler. The market itself usually comprises a number of stalls or 'stands' belonging to or rented by particular wholesalers who buy and sell on commission. On arrival samples of produce are commonly displayed so that buyers can view the range available. The salesman will have a good notion of prevailing prices and of the state of supply and demand. He will also have close and often long established contacts with the buyers. The asking price will be increased or decreased according to fluctuations in supply and demand during the period of trading. This inevitably leads to a degree of 'horse-trading' and value judgement on the part of salesmen but this may determine whether or not the produce actually reaches a consumer. In the case of highly perishable produce in a sudden period of bad weather or towards the end of the morning trading period, making a sale at any price may be the overriding feature. In reality the system can be highly sensitive to change in market conditions.

Primary wholesalers subtract a commission from the total sale of the produce so it is in their interest to sell for the highest price possible. The secondary wholesaler who actually buys the produce 'marks up'. Wholesalers are in competition with each other for the total volume of business, a further safeguard of efficiency. Growers themselves and other merchants may also be granted selling space in some markets. They sell to wholesalers or retailers.

Common problems of wholesale markets

The longer established markets are often located at the centre of now large conurbations. The pressures brought by the expansion of the population, urban development and the motor car have led to congestion in and around the sites. Lack of mechanical handling facilities and inaccessibility to buyers often lower the efficiency with which produce can be handled. Many markets have been rebuilt using more spacious sites on the outskirts of cities where they are usually better served by motorways and arterial roads. Well known examples are the replacement of the famous Les Halles market by the Paris-Rungis wholesaling centre south of the city and the current reconstruction of Covent Garden market at Nine Elms.

It is natural that criticism should be made of the distributive industries by producers for they take a large proportion of the final price. That profits are too high and that too many people are involved in marketing are commonly voiced objections. Four separate investigations in the UK between 1920 and 1968 failed to identify excessive profits in horticultural wholesaling; on the contrary, they appeared to be quite reasonable. Intense competition within the market, particularly with diminishing total trade and a decline in the number of wholesalers, ensures hard bargaining.

Complexity of the market network is common as is evident from the diagram of the routes by which produce reaches consumers in the UK (appendix 1). Its very complexity does however ensure that produce reaches the consumer; it has flexibility. Attempts at simplification and shortening the chain all bring further problems. By-passing the markets by coordination of retailer-grower or processor-grower usually requires some form of contractual arrangement, which is discussed later in the chapter. Simplifying the market itself by the use of a clock auction is only successful when particular conditions exist (see page 45). Vital and closely related factors in efficient wholesale trading are market intelligence and standardisation. Growers and buyers must have market information before they can plan their activities and they are handicapped always because all market intelligence data are historical by the time they are available as planning factors. The telephone is in practice the most useful medium of communication of market intelligence. Market intelligence when on hand is useful only if standard factors of type, form and quality of produce are available for reciprocal appraisal; market intelligence alone is valueless without these essential complements. In the market, buyers are able to purchase confidently, not needing to inspect every unit of produce if they can rely on standards. A working party of the United Nations put forward recommendations in 1955 for many horticultural products. These now provide the basis of several national and international standards taking the place of a multiplicity of different ones in different countries. Agreement on common standards is imperative when produce crosses international frontiers.

The inability of markets to cope with gluts and shortages is a common criticism. It is further claimed that margins are inflexible ('sticky margins') when changes in supply occur. Imperfections in the market in the short term are to be expected. A salesman may, for example, receive strawberries at the market in a period of hitherto warm dry weather. On the morning it may be cold and wet, deterring buyers and lowering the overall demand and price for strawberries. If the rain starts during trading the salesman may make a quick sale, even at a lower price, knowing that the strawberries will not store until the next day. Such variation in relation to the weather, size of consignment, type of package has been identified in Birmingham market in the UK. The author has further recorded prices varying by a factor of three in the East Midland wholesale markets for the same type of chrysanthemums on the same day.

The 'sticky margin' accusation leads to complaints that profit margins are maintained even though in general prices have fallen. Substantiation of this charge is hard to establish, but there is some evidence from Holland, Italy and the UK for potatoes in particular. The principle is best illustrated by an example

Buying price	Selling price	Margin
p	p	p
10	12	2
8	10	2

The wholesaler has retained the same margin in both instances but both the buying and selling prices have fallen.

Too much reconsignment of produce is another common criticism levelled at wholesale traders. This occurs mainly between primary and secondary stages. It arises partly from the general distribution arrangements allowing produce from all over the world, potentially at least, to find its way into all retail shops. It also occurs as a means of alleviating local shortages and gluts. Deficient markets in terms of particular products can quickly make their needs known by telephone to markets with plentiful supplies. The higher prices pertaining as a result of shortage are often adequate attraction. Reconsignment is also common when there are too few market centres in relation to the geographical distribution of consumers. Rapid effective transport is essential for reconsignment which is limited if the transport network is inflexible in any way. The closure of a branch railway line or the opening of a new motorway can radically change the efficiency with which a district can be supplied.

(b) Auction markets

Auctions are private or public markets where products are sold to the highest bidder. This type of market exists for horticultural produce but another type, the Dutch Auction, has become particularly significant. It involves the downward movement of prices, the first acceptor being the recipient. This system has been adopted in many Western European markets since its inception in Holland at the beginning of the century. To facilitate sales an electric clock indicator is used. It is set in front of the buyers who record their bids from desks arranged like a lecture theatre facing the clock. The sequence of activities might be as follows. Potential buyers inspect the produce labelled with lot numbers before the sale commences. As the lot is announced the auctioneer sets the clock moving downwards from a price that he considers well in excess of the expected price. The buyers, wholesalers, retailers and exporters, watch the clock until it reaches a price at which one of them is prepared to pay. He presses the button on his desk, the clock stops and his purchase is recorded usually by an illuminated number in the face of the clock and also by the auctioneer's clerks. In modern premises connection to a computer can facilitate automatic recording and invoicing of the purchase. The buyers are in competition; to purchase they must press their button before their colleagues, and because the device is electronic one buyer is always first.

The advantages claimed for the Dutch Auction are fourfold: greater speed of selling; greater accuracy with the electronic clock eliminating recording errors; better prices for the grower that are truer reflections of prevailing supply and demand; and greater honesty in trading. A common criticism of conventional auctions has always been that 'rings' of buyers may get together to deliberately withhold bids to keep prices low. Another practice is to use fictitious bids to raise prices artificially.

To be successful the Dutch Auction requires a high proportion of standardised produce, in large quantities, implying a high degree of specialisation of production and bulking of small growers' output. The buyers themselves must be accustomed and even trained to the use of the clock. The importance of these factors is nowhere better illustrated than in the failure to establish a clock auction at Cheltenham in the UK. The clock was introduced to a conventional auction market where buyers were used to ascending prices. Cheltenham is not central within a concentrated area of specific horticultural production. It is close to an area in the hands of small-scale producers and grading standards at that time were not high and were loosely interpreted. Low prices were generated, buyers and authorities disagreed, consignments of produce to the auction were reduced and after spending a considerable sum of public money and enduring mechanical problems with the clock itself its use was abandoned after nine months.

In contrast, production in the main horticultural areas of Holland, Belgium and parts of Germany are very concentrated and here the clock has been successful. Furthermore buyers and producers seem better disciplined in accepting grading standards and the workings of the auction.

3.2 GROWER ORGANISATIONS

(a) Marketing boards

A marketing board is an example of producer horizontal integration usually with Government sanction and compulsory membership. Producers generally have a predominant membership of the board. These features may give the board virtually monopoly powers in the market. Erosion of this power due to poor statutory organisation or disobedience on the part of members can be a major problem affecting success.

A basic aim of the board may be to raise the long term incomes of members by exercising monopoly control over supply to maintain prices at some predetermined level or minimum. It may also attempt to manipulate demand by advertising on behalf of producers. Research into new markets and aspects of production may be financed and coordinated by the board. The monopoly power is further used to counterbalance monopsonist tendencies on the part of buyers in the markets. This is called counterveiling power.

Supply control is by far the most important function. It may be achieved either by controlling the amount of crop grown, the area sown or planted, or most commonly, by controlling the amount supplied to the market at a given time. Produce may be diverted from the market to alternative uses and even purchased from the market with centralised funds.

The main advantages of producer marketing boards are firstly they may provide a strong economic incentive to profit by ensuring satisfactory and stable prices to producers. This leads secondly, to market security for growers who are then able to consolidate their businesses. Thirdly, this type of income support can often cost less than direct financial assistance by the Government.

A major disadvantage is the artificially raised prices, this burden being passed on to retail consumers so the poorest section of the community suffers most. Land itself may artificially increase in price particularly if supply control is by acreage quota. When quotas are limited to specific pieces of land trade in quotas often develops. This further leads to an ossification of the areas of production of particular crops. It may even keep inefficient producers and systems going at the expense of more efficient ones. Furthermore the need to supervise acreage restrictions and to collect levies can result in the evolution of a costly and bureaucratic organisation.

An economic disadvantage arises from any interference with the supply of a product. It automatically alters the relationship of the product with its substitutes. For example, if the price of apples could be increased, consumers may well transfer their purchases to oranges. Effective supply control assumes considerable knowledge of the nature of the demand curve. This poses the question – will supply control raise the price by the same amount at all points on the demand curve? Consider for example, the hypothetical situation depicted in figure 3.2. If supply is withheld in relation to the demand curve DD at q_1 a price p_1 can be expected. As the quantity supplied is further controlled at say q_2, q_3 and q_4 the rise in price increases at each point. If the demand curve were perfectly elastic (horizontal) there would be no price response over that range. Lastly society is very wary of vesting monopoly in any organisation because it might be abused.

Marketing boards are considered further in chapter 8 as tools of Government policies.

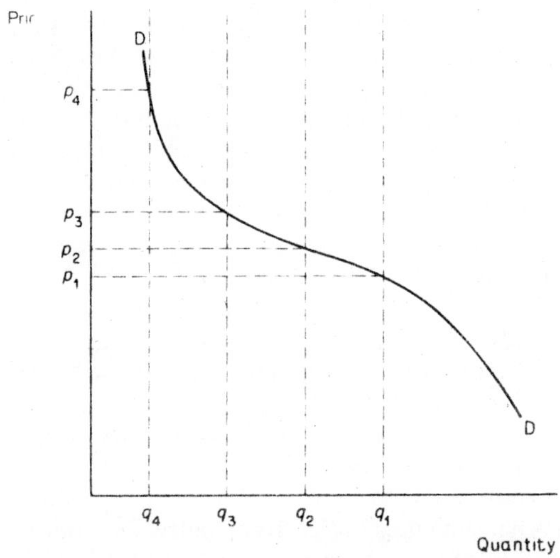

Figure 3.2

(b) Cooperatives

Cooperatives are another form of horizontal integration. Cooperation might be defined as any joint action for natural benefit but *a cooperative* in horticulture may refer to a formal organisation or company. In most countries it involves a union of growers to provide services for members on sound economic lines aiming to avoid loss and having open membership. Control is very democratic to the extent of one man, one vote. Capital is raised from members and surpluses from trading may be redistributed according to patronage. The main services provided are marketing, bulk buying, technical assistance and credit, but other forms of cooperation exist. Our concern here is with marketing.

Marketing cooperatives may offer services as follows. Firstly by acting as negotiators with retailers and processers to secure better prices and contracts. Secondly by integrating with other cooperatives or large growers to form so called 'groups'. This is not to be confused with amalgamation or take-overs but refers to a lateral coordination of marketing activity. Individual identity is not lost even though the produce may be all marketed under one brand label. It is sometimes called second tier cooperation. Thirdly they may offer a marketing service to individual members enabling them to shed the task of marketing and concentrate on production. The grower pays for the service by shares in the organisation and additional levies are usually made according to quantity of crop handled. These may be simply deducted from market returns. Lastly the cooperative may integrate vertically and invest in processing and even retailing facilities.

Many of our daily activities are undertaken in groups. Group theory may form a very large part of a sociologist's studies and factors which make for strengths and weaknesses of groups have been studied. The principles are equally applicable to groups of growers in collective action.

Consider a successful association football team. It derives its successes from the overall skills of its members and their ability to play as a team. The manager is a key figure in choosing players, determining strategy and delegating responsibility by choosing a captain from the players to maintain team spirit, discipline and encouragement. Management is also responsible for recruitment and training. Players and management communicate freely and have a clear objective of goals and to win. All these features apply to horticultural cooperatives if they are to be successful.

Unsuccessful teams, by contrast, may have poor players and individualism commonly causes disruption and poor team performance. Management control is generally poor in team selection, strategy and recruitment. The club may suffer from petty bickering and jealousies. Objectives may be reduced to mere survival in the league. These features are all too common in horticultural cooperatives that fail. Starting a cooperative is not easy. Once started, maintaining the growth and progress of the organisation may present difficulties.

Some common problems of cooperative development are enumerated below

1. An inherent weakness is the very basic cooperative principle of one man, one vote. Grower members may not necessarily be as good at

marketing as they are at growing. Over attention to democracy in the decision making meetings may prevent the cooperative manager from taking the decisions that are vital to ensure long term success.

2. Often only a single share is required to become a member. This leads to a further common weakness, that of capital starvation. Members may be reluctant to invest more monies in the cooperative or to accept schemes to accumulate capital from cooperative sales.

3. Managers are often undervalued and even under used. Low salaries have commonly been paid to managers of cooperatives, thus affecting recruitment of men of the right calibre.

4. 'Top heaviness' is a common criticism made of some cooperatives. The hierarchic structure of management directors and members in some cases has caused communication difficulties.

5. Communication is an essential feature of any successful management. The very nature of horticultural business means that members are geographically dispersed from the central cooperative headquarters. It is imperative that communication be two-way between members and the cooperative management. It is particularly important to ensure that the overall objective of the company in terms of crops and quality required by the market is understood by the members. All too often growers are not critical enough or adequately informed or even interested in the central activities.

6. Class, social and political aspects of particular countries may determine the degree of success of cooperation. In Denmark and Holland there appears to be considerable readiness on the part of farmers and growers to cooperate. In the UK independence seems to be valued more highly. Political neutrality has always been claimed by cooperative organisations in the UK but this image has suffered and still does by the apparent association of cooperation with the Labour Party from 1917 to 1926.

7. There is a common and understandable reluctance to form larger units either by increasing sales, taking on new members, amalgamation, take-over or merging. Successful cooperatives of a particular size may not achieve greater profitability merely by increasing size, especially if this changes the nature of the organisation. We understand all too little of the sociology of cooperatives.

3.3 RETAILING

A wide variety of retail shops sell horticultural produce from 'barrow boys' to large multiple supermarkets. In the UK there are probably about 150 000 shops selling commercial horticultural produce, about one per 400 consumers. Only about a third of these are sufficiently specialised to call themselves greengrocers, fruiterers or florists.

The dominant feature of retailing in the developed countries during the past twenty years has been the growth of self-service purchasing. This has brought with it a need for high quality prepacked produce assembled in large quantities and regularly supplied. The implications of this for horticultural producers have been enormous. Growers have been forced to programme their cropping to satisfy these outlets in terms of quality, continuity and size of consignment.

On the High Street the independent greengrocers and fruiterers have lost ground both in sales and number of shops as the large multiples have expanded. As with wholesalers there is little evidence to suggest excessive profits amongst retailers of horticultural produce. The fact that numbers are falling is further evidence for this.

'Mark-ups' on purchases by retailers are often high but not exceptionally above the average for other goods. The trading risk is considerable and margins and profitability are very similar to those for fish, meat and poultry which have similar characteristics. Florists have the highest of all retail 'mark-ups' and the 'sticky margin' criticism is levelled at retailers as well as wholesalers, for it is widely held that retailers keep prices high when growers' prices fall. Retailers would reply that it is not easy to increase their turnover when glut conditions exist and that they are in competition with other retailers so they cannot keep their prices consistently higher than their competitors.

In recent years small-scale retailers have become better organised. Many have introduced self-service as a means of winning back customers from the new competition. Notable in this field are the so called voluntary chain stores and in the UK the Cooperative Society shops. Voluntary chains such as 'Vivo', 'Spar' and 'Mace' are collective organisations of small scale retailers grouped for bulk buying and distribution. The trade name is used in common and some of them already operate throughout Western Europe. The trend towards self-service and cooperation varies enormously from country to country but in Western Germany about 40 per cent of all retail traders *buy* cooperatively.

The motor car era, particularly in developed countries, has brought a new dimension to retailing. Mobility of consumers has led to two important developments. Firstly the evolution of suburban shopping centres concurrent with a trend towards decentralisation of towns and cities for dwelling purposes. Commuter suburban dwelling is on the increase. The new shopping centres are often more spacious and operate at lower cost. In the USA and France very large 'Hypermarkets' have appeared selling a very wide range of products often in bulk packs. The general increase in number of home freezers together with the possession of a means of convenient transport has aided the development of bulk sales generally.

A second development that the motor car has brought is the entry of farmers and growers into retail trading. Road-side stalls have appeared all over the countryside. To the grower this means the elimination of the middleman but as many have found it is not without problems. Often a large range of produce has to be stocked and to comply with local by-laws regarding parking and hygiene of retailing premises can often be inhibiting or too costly. The display skills of the retailer do not always come easily to the grower and he quickly finds that there is an upper limit to the amount of produce that can be sold in this way.

Institutional consumption

We generally think of horticultural marketing as supplying food and decoration at home. A neglected but increasingly important outlet in the higher income countries is institutional or catering consumption. This includes hotels, restaurants, cafes, outside caterers, factory canteens, schools, universities, colleges, hospitals, military establishments and travel catering. The tendency to 'eat out' increases as incomes increase. In the UK this trade is probably worth at least £1000M at the point of sale, about 60 per cent being the value of food and the remainder largely services. This is about 10 per cent of the total value of food consumption.

Much of this food is purchased by the caterers in convenience form because of continually rising labour costs for preparing meals. In general the caterers' requirements are the same as those of other large scale buyers – continuity, quality and a wide range of produce.

Contractual arrangements

An essential result of the coordination of marketing of producers, processers or retailers, is the introduction of some form of contractual agreement. This is necessary for the processor who needs to ensure supplies of raw material to keep his processing plant working. Similarly the large scale retailer needs to fill his selling space regularly to maintain turnover. For the grower it is a way of securing certainty of a sale of his produce in the future and possibly a price.

The general situation is one of many growers bargaining with fewer buyers – imperfect competition. The few buyers are often better organised, having superior information and ability for bargaining and decision making. Often their businesses are already diversified into several activities. Most processors distribute a wide range of convenience foods and often estimate and influence the market for their product. In the USA in particular growers have formed Bargaining Associations to negotiate with processors. These organisations must control the whole or preferably a large proportion of supply otherwise the processors may still be able to purchase more cheaply elsewhere.

One of the main purposes of a contract is to eliminate as far as possible large fluctuations in price. It may also attempt to establish a fair price for both parties in the longer term. A greater certainty of the market permits planning for the future, although there can never be certainty for any length of time. Some inescapable variations are inevitable because of human errors and the nature of horticultural production. There are two central difficulties that must be conquered. Firstly to arrive at a long term bargain which does not cause firms to suffer in competition. This is particularly important when intense competition exists between processors at the point of retailing. Secondly to ensure that the good will of growers is not lost through setting the price too low, thus threatening supplies in the future.

The types of contract generally employed are

1. A *fixed price contract* – an agreed formula of prices for the future as assurance for the grower. The risk involved is passed to the customer. When the fixed price becomes out of line with market prices supplies may become excessive or in deficit according to the disparity.
2. A *cost plus contract* – the cost of production plus a margin for profits, providing it is accurate and reviewed constantly, provides reasonable future certainty.
3. A *market price plus contract* – leaves the grower exposed to the whims of the market but does give him benefits for producing superior quality.
4. A *no price contract* – apart from a guarantee to purchase no price is agreed; this may increase the risk to the grower. The agreement negotiated will fit into one of the above categories but will contain greater detail.

Retail buyers generally prefer to maintain a 'gentleman's agreement' with suppliers. They have only a verbal agreement or letter of intent to take a particular quantity of produce at a particular time. It is in the interests of both parties to maintain contact and exchange but there is an escape mechanism for the grower if the requirements suddenly cannot be met. This might be due to a hailstorm or other extreme weather hazard or even bad planning.

The degree of integration between the grower and the contracting buyer may not be confined to marketing but sometimes extends to the provision of 'fieldsmen' employed by the buyer to give technical assistance and even financial assistance may be provided on reasonable terms.

CHAPTER 4

PRODUCTION

We have considered the market for our produce so we can now begin to examine the economic characteristics and behaviour of the producer. In the first chapter it was shown how horticulture emerged as an industry in economic development. An industry is a collection of producers or firms and 'industrial structure' refers to the collective characteristics of these firms.

4.1 THE FIRM

Economists call this part of economic analysis the theory of the firm. A firm is defined as an organisation to provide and allocate limited resources in such a way as to maximise profits. The resources are those of capital, land and labour which are combined in particular proportions to produce output or horticultural produce. Note that this definition also stipulates maximum profit as a goal of the firm. This as we shall see later is not necessarily true in the real world but is often taken as a standard by economists.

An accountant, by contrast, might simply see a firm as an organisation having an identifiable collection of assets and liabilities with funds available or owing. In turn a legal definition might be embodied in such phrases as 'a sole proprietorship', 'a partnership' or 'a company with limited liabilities'. These are all specific definitions and all refer to business organisations. All have one very important common feature, they are engaged in productive activity – the transforming of inputs into output. This comes about as a result of decisions made by the controller of the firm – the manager or entrepreneur. This latter word refers specifically to the person who is responsible for organising the resources.

We can now define the manager as the person who organises the resources of the firm to maximise profits. It is another formal definition which differs from more popular definitions that often refer collectively to an organisational structure, *the* management. Many, particularly non-agricultural firms, now have multi-management structures with a hierarchy of decision making from a managing director and board of directors to plant managers, foremen, chargehands and so on down the line.

Agricultural industries are characterised as comprising a large number of individual holdings. Throughout the world agricultural holdings are of the traditional, single, entrepreneur type, the farmer not only being the key decision maker and capitalist but also part of the labour force. It is only in recent years that larger multi-management firms often employing graduates have emerged in the horticulture of Western Europe and North America. Most of the world's farmers are peasants; according to the FAO over half of the world's population is engaged in agriculture or dependent upon it for a living. Peasant farming however can be highly organised and this description should not be seen as a judgement of inferiority. Japan, Burma and Thailand are examples of countries having considerable technical competence, the last two being major exporters of rice from peasant holdings.

Information regarding the structure of horticultural holdings in particular is not internationally available. It can be safely assumed that most horticultural producers still conform to our entrepreneural definition. In the UK where differing estimates of the number of horticultural holdings have been made, it is likely that there are at least 55 000 holdings which with total sales of about £200M means that the average holding has sales of only £3 600. Most of the production is in the hands of the larger producers, so the UK must have many very small scale growers.

The grower

On most holdings the horticultural producer or grower is also the manager. His holding, usually on good quality land, may be isolated but is often adjacent to holdings of similar producers. The rural nature of the business means that his decision making is done in relative isolation compared with other more concentrated industries, so self-confidence is a useful attribute for the grower.

The fortunes of the business are very dependent upon the prevailing climate. Often production is highly seasonal with physiological limits being imposed on what may be produced at a given time and in a particular place unless recourse is made to capital investment in storage or protective covering for the crops. The produce, often highly perishable, may travel long distances and this demands considerable expertise and organisation on the part of the manager. Indeed the attributes of a successful grower are numerous, covering a wide spectrum of subjects. He may need to be an engineer, botanist, plant pathologist, technician and business man all at the same time.

The profit motive

The word profit has many different everyday meanings. In economic analysis *pure profit* is the difference between the sales of the firm's output and the cost of the inputs required for its production. These inputs are costed by comparing them with their best alternative use outside the business (or their opportunity cost – see page 87). *Normal profit* is that level of profits which will just keep the firm in the industry in the long term or for more than one production season. These are formal definitions of pure economics. The reader may also encounter, particularly in commerce, expressions such as 'accounting profit', 'taxable profit' and 'net farm income'. These have specific meanings according to the purpose for which they are considered. For example, accounting profit usually takes into account wages that the owner-manager might earn by selling his services and the interest on investing his money elsewhere.

Economists use the maximum profit motive in the same manner as irrigation users talk of 'field capacity'. It is a realistic base point from which to measure. Here the analogy ends, for no attempt is made to measure regimes between nil and maximum profits. In practice firms do not always seek maximum profits. Apart from the obvious problem of identifying when this point is reached it is very common to hear growers saying 'of course we aim for the highest profit' or it is implied in 'we aim to make as much money as possible'.

Almost everyone has particular fads and fancies making for irrational decisions which considered logically lead to less than maximum profits. It is probably better to talk of the goals of a particular organisation and to define them in terms of the prevailing situation. For example, a grower may wish his son to follow him in the family business and therefore must generate sufficient profit to pay for his education at school and possibly university. He may also like to go sailing and play golf and attend markets and political meetings regularly. The actual goals set will vary considerably between producers according to their aspirations. The provision of sufficient food for survival may be the main pre-occupation at one end of the scale. At the other end profit may be defined at a specific acceptable level with provision for recreation and even part-time business activity.

It is important to understand the relationship between inputs and outputs and how profits arise. Figure 4.1 illustrates for a single period the relationship between a firm's total revenue TR and total cost TC. Costs and revenue each increase with additional output and may have the general form depicted. The nature of the curves is explored in greater detail in chapter 6. Profits arise when the TR curve exceeds the TC curve between Q_1 and Q_2. At Q_3 maximum profits occur when the vertical difference between TR and TC is greatest. Note the consequences for a firm whose goal is to maximise sales. It may find itself sacrificing profits. For various reasons there are firms that do not produce profits at all.

4.2 INPUT – OUTPUT RELATIONSHIPS

In figure 4.1 the inputs are represented by the total cost curve and output by the total revenue line. The quantity of output depends upon the quantities of inputs used in production. The relationship between them is called the *production function*. It can be defined mathematically and this can become exceedingly complex but analysis to this depth will not

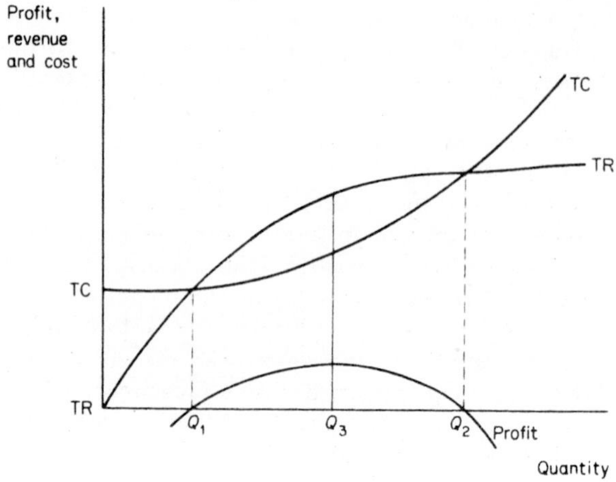

Figure 4.1

concern us here. In the simplest form we could write a production function as $Y_1 = f(X_1)$ where Y_1 = output and X_1 = an input. This tells us that the quantity of Y_1 produced is a function of the amount of X_1 used in its production. It could be yield of carrots (Y_1) in relation to the amount of seed (X_1) used. Clearly to be of any great value quantitative information is necessary. The producer needs to know how much seed will give him a particular yield. He must also be aware of all the other inputs and factors affecting the production of carrots. If all the inputs are included in the equation we can write it in a generalised manner as $Y = f(X_1, X_2, X_3 .... X)$. In this example we kept all the other factors steady and only considered variations in the use of X_1, the seed rate. The same procedure is normally adopted in experimentation and the greater the number of variables considered at one time the more difficult it is to reach conclusive results.

If the relationship between Y_1 and X_1 takes the form of a straight line when plotted graphically there are constant returns of Y_1 from the use of X_1. This is a very simple but rare situation in reality. If the returns are increasing the curve is upward sloping and if diminishing downward. These are three very simple forms of production functions that may occur when a single input is varied. These are illustrated in figure 4.2 as a curve TP combining all three which may occur in reality.

Regrettably, due to lack of funds and little cooperation between natural scientists and agricultural economists, there is a dearth of data regarding input/output relationships in horticulture. For agriculture and animal feeding in particular the situation is somewhat better particularly in the USA. As it becomes progressively more difficult to raise yields profitably it will become more and more in the interests of both scientists and growers to cooperate towards this end.

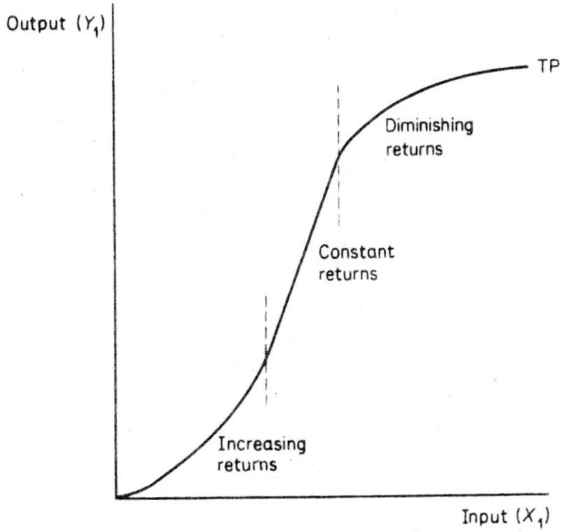

Figure 4.2

The law of diminishing returns

The diminishing returns depicted in figure 4.2 are inevitable after a certain quantity of inputs is used. This fact has led to the proposal of the so called law of diminishing returns or law of variable proportions which may be defined as follows: if successive units of an input are added in conjunction with a fixed stock of others the increments will after a point yield successively smaller and smaller additions to the total output.

Students of horticulture will undoubtedly have encountered probably the best known example of this law, the response curve. Examples might include responses to fertilisers, seed and plant densities. The curve in figure 4.3 represents a typical but hypothetical response curve showing diminishing returns. We can analyse this curve further to obtain a better understanding of the production function.

Average, total and marginal products

The input X_1 in figure 4.3 increases in equal increments but the output Y_1 declines for each successive input increase giving true diminishing returns. The diminishing outputs are labelled *a b c d e* and *f*. The change in output for a single increment of input is called the *marginal product* (MP). It is the rate of change of Y_1 in relation to X_1 or $\Delta Y_1/\Delta X_1$. Since it refers to physical data it is called the marginal physical product (MPP). If the price to be received for the product is known by multiplying the physical product by the price, the *marginal revenue* (MR) results (Revenue = Yield x Price). The marginal concept is extremely important in economics. We introduced it in chapter 2 and it will appear again in chapter 6.

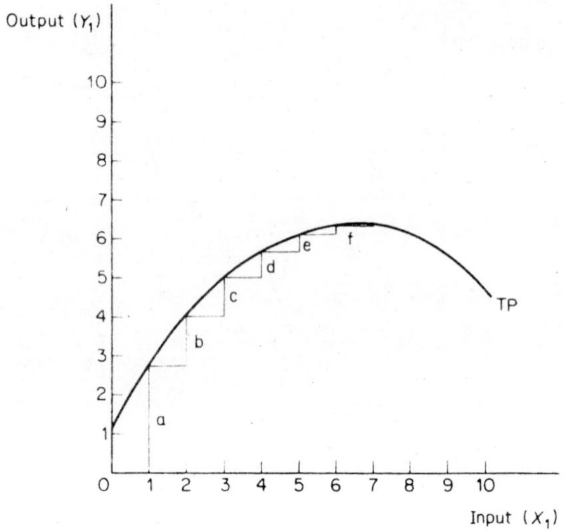

Figure 4.3

In the same way as a cricketer's average score is calculated by dividing the total scored over a period of time by the number of innings, so it is often useful to determine the *average product* from using equal increments of inputs.

It is the ratio of total product to inputs or Y_1/X_1. The value of the average product shows diminishing returns to input after passing its maximum point.

Table 4.1

Average and marginal products

Input (X_1)	*Output* (Y_1)	*Average product* (Y_1/X_1)	*Marginal product* ($\Delta Y_1/\Delta X_1$)
0	0	–	
			2
1	2	2.0	
			4
2	6	3.0	
			7
3	13	4.3	
			7
4	20	5.0	
			3
5	23	4.6	
			1
6	24	4.0	
			−1
7	23	3.2	

Using some hypothetical data (table 4.1) we can construct a diagram (figure 4.4) to illustrate these expressions and the relationship that exists between them in the production function.

Both the AP and MP curves are derived from the TP curve. The graph has been deliberately drawn in two parts so that the Y_1 axis could be enlarged to make the presentation lucid. Note also that the marginal product has been plotted mid-way between the increments of X_1. It is also presented in this way in table 4.1 because our calculations of marginal products are themselves averages for each increment. The student should check the calculations of average and marginal products in table 4.1 and try plotting the graphs for himself. This will not only assist in following the reasoning but will help the student toward a better understanding of the relationships.

Rational and irrational production

Figure 4.4 has been divided into three distinct stages. The grower's concern is to decide at what stage of the production function he should aim to gain optimal output from the inputs that he uses. In this respect stages I and II of the model can be eliminated as irrational for reasons that we shall now discuss.

Stage I continues to the point where the average product reaches a maximum but the total product is still increasing. The marginal product has already passed its maximum and as it falls cuts the average product curve at its maximum. Total product is always increasing within this stage; the more input used the greater will be the output obtained. It follows therefore that it is irrational for a true profit maximising grower to limit production to this stage.

It will always pay him to keep adding inputs whilst the average product is increasing. Examples of this type of irrational behaviour might include insufficient use of fertilisers or lime because of lack of experience or information. A considerable amount of research and even cash incentives are directed towards this, as is exhortation to farmers and growers to move beyond this first stage in this particular use of resources.

Stage II – total product increases to a maximum although the average product is falling. The marginal product falls to zero. Within this stage even though diminishing returns are evident there is still an increase in total product for increasing input. It is within this stage that the most profitable level of production occurs; it is the region of rational production. From physical data alone it is only possible to identify the range of most profitable production. We need financial information to be able to specify precisely what level of inputs the grower should use. We shall do this in chapter 6.

Stage III – any increase in the use of inputs in this stage results in a falling total product and the marginal product becomes negative. This stage is therefore irrational. Growers may well get themselves into this stage

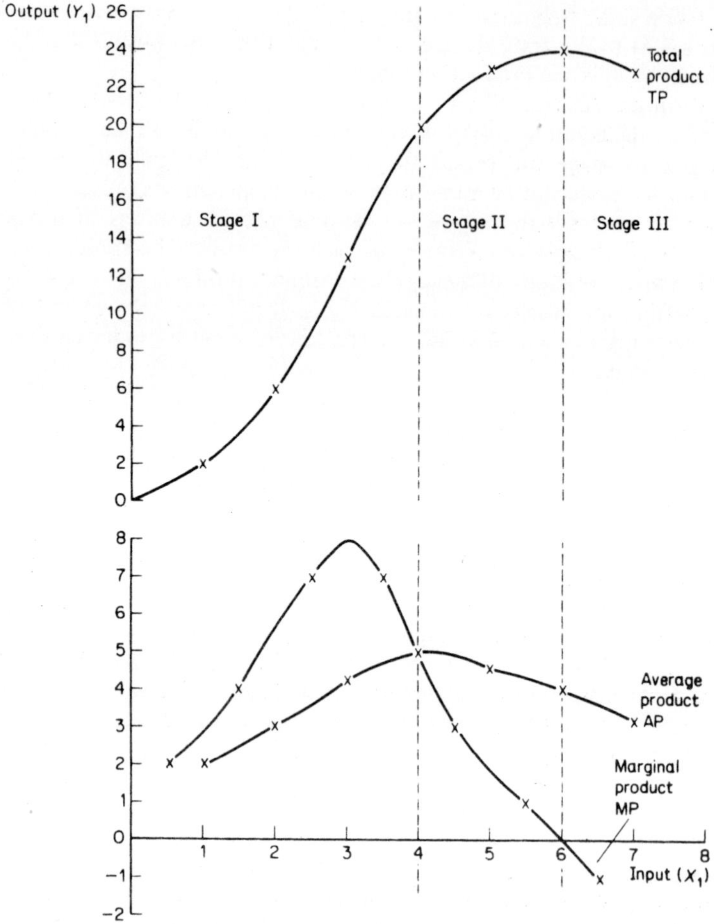

Figure 4.4

usually in the absence of knowledge or experience or even because of carelessness. They may use too much nitrogenous fertiliser, for example, and suffer losses from disease or produce excessive vegetative growth at the expense of fruit. Another example might be the result of a too dense plant population resulting in excessive competition between plants at the expense of yield.

The simple production functions that we have considered so far concern one product and a single input varied in a particular way over a specific time period. The use of more than one input and combination of outputs is considered in chapter 6. It is very important to define the time period involved in all economic analyses. A production function established in one month may be quite different in another, due to changes in the other inputs used and the method of their combination. This may be particularly important in countries having a marked seasonality. For example manpower may have equal productivity all the year round in a central African country but in North America outdoor work may only be possible for half the year because of snow and frozen soil.

Technological advance

If the quantity of output obtainable from a given level of input is increased a technological advance has occurred. This might result from the breeding of a new cultivar superior to existing ones or from the use of an improved type of fertiliser. This might be represented graphically as in figure 4.5 which depicts the two production functions. The cost of using the new cultivar must obviously be considered but if there is no difference in cost then the improvement is clearly superior.

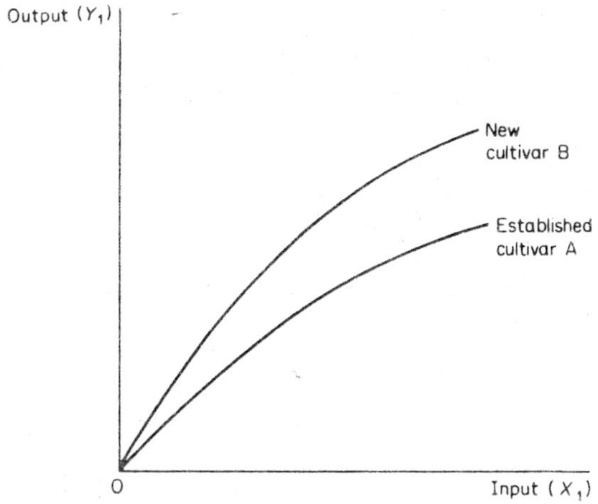

Figure 4.5

So far in this chapter we have considered the nature of the firm and the characteristics of horticultural producers. We have also introduced the concept of input and output relationships. Before we can consider further the use of resources used in production we must first deliberate on perhaps the most important economic feature impinging upon horticultural production. Events occur in time but we are unable to predict precisely events of the future. We say that there is uncertainty about the outcome of future actions. All production involves a degree of uncertainty but characteristically horticulture has a particularly high level. It affects management and marketing activities alike and is a key feature in all business organisations and activities.

4.3 RISK AND UNCERTAINTY

Although often used interchangeably risk and uncertainty have different meanings to an economist. Risk is predictable, uncertainty is not. The number of fires on farms or hailstorms occurring within a given area and time can be recorded and a reasonably accurate prediction made as to their likely occurrence in the future. We cannot of course predict exactly where they will occur. Insurance companies make calculations of this type in assessing the premium payments necessary when issuing insurance certificates. Yield variability due to weather, pests and diseases cannot be predicted in this way, nor can the activities of other people and business associates. A grower, for example, may be 'let down' by a machinery supplier at a critical time during the crop harvest. Throughout the remainder of this book we shall refer only to uncertainty as a generalised expression including risk.

Profits, management and uncertainty

If the future were predictable, the grower knowing in advance what yields he was going to achieve and the prices that he would receive from the market, all that would be necessary is for the manager to set in motion a production plan and supervise its progress. Thus management, except for setting the initial plan, would be eliminated. Furthermore, all firms in the industry knowing the future to be certain would equate their marginal revenue and costs (see page 93) and in conditions of pure competition there would be no profit. Profits and the need for management arise from uncertainty.

In the real world there is uncertainty so that there is a need for management and it follows that profits are inevitable to some while others make a loss. The burden of uncertainty has to be borne by someone and it is the lure of potential profits that ensures the acceptance of this burden. Characteristically the highest profits are generally associated with the most uncertain situations. Under such circumstances the choice facing the manager may well be between profits at a high but uncertain and fluctuating level or lower but more stable ones.

Profits realised by management are partly the result of good judgement about the future and partly chance. The function of management can be set down in the light of uncertainty as follows

1. To make predictions about future business activities
2. To make plans consistent with these predictions
3. To ensure that the plans are carried out
4. To stand the consequences of the outcome

The last function reveals to the manager whether or not further uncertainty is worth bearing. He may continue with the plan or even further commit himself if the outcome is attractive. He may well modify his business plans to ward off some of the uncertainty.

Types of uncertainty

The first category of uncertainties are those due to natural phenomena. Variability in horticulture due to the weather is an obvious and inevitable feature. The actual incidence of variation will depend upon the location and type of business. The effects of flooding, drought, hail, tornadoes, hurricanes, lightning and freezing on crop yields are well known. The weather also affects the ecology of others including pests and pathogens, the incidence of which may be advanced or retarded according to prevailing conditions. Insects are often quoted as consuming 14 per cent of world crops and even as high as 21 per cent of tropical crops in Asia.

Social problems are a second category of uncertainties. They include civil riot and even war, changes in social structure, strikes, theft and embezzlement. Strikes are not a common feature of agriculture but the closer it aligns itself to other industries the greater the chance of them becoming a problem. A strike in another industry supplying vital materials to the grower could be serious. If by market porters, for example, it would affect prices and the means of selling. Civil riot increases the risk of death and injury and damage to property. Growers in Northern Ireland in 1971 found that civil strife in Belfast, the main market and consuming centre for their produce, caused considerable hardships. Not only were some retail shops bombed but some growers' vehicles were destroyed, acting as a deterrent to others to send to market. A change in social structure from say subsistence to a commercial exchange economy brings its own problems for horticulture if only in terms of keeping abreast of technology. Advances in technology may also bring uncertainties in the form of a higher accident potential associated with the use of more complex machinery.

Thirdly, there are economic uncertainties which afflict the grower. The main one is that arising from fluctuating market prices not anticipated by the grower. The relatively inelastic supply and demand for horticultural products means that small changes in either lead to large fluctuations in price. It is often not realised just how sensitive the market is to such changes. Prices generated in wholesale markets can be very variable indeed. The author recorded the following range of prices for chrysanthemum blooms of similar quality, all realised in one morning in the wholesale markets of the East Midlands of the UK. Six different growers received 36½p, 70p, 50p, 30p, 80p and 45½p per twelve blooms.

Further economic uncertainties arise from the process of increasing capitalisation. Large expensive investments increase uncertainty because the rate of obsolescence may be accelerated by rapidly changing technology.

Fourthly and lastly, we can identify personal uncertainties. These involve the human factor and may affect the grower and his workers as well as third parties. Death, injury, illness and maternity are obvious examples. A grower dying early and leaving an encumbered estate may cause his family considerable financial stress. He may alternatively survive to a ripe old age but be incapacitated as a worker. He might find himself held liable for accidents suffered by his employees or third parties, for high accident rates are a common feature of agricultural work. Several reasons can be put forward to account for this. Complex machinery used in isolated and sometimes inadequately supervised situations can cause many accidents. First aid may not be available and adequate up to date training in its application is seldom given to agricultural workers. The average age of such workers is often high and long hours are worked at peak times. In the USA fatalities have not decreased with the declining labour force and are estimated at an annual rate of some 60 to 70 per 100 000 of the farm population.

Combatting uncertainty

Whether or not the grower is concerned to try and reduce uncertainty will depend upon his particular goals and availability of capital. If capital is plentiful and his debts few or non existent then he may choose the more uncertain activities. If a new grower adopting the same plan has a bad first year it may prove disastrous. Precautionary measures to combat uncertainty can only be used *at cost* to the operator either in lower output or increased actual cost. The main categories of measures available to management are considered here.

1. *Insurances*

To be insurable a phenomenon must have some degree of predictability. This further implies facilities for the collection and analysis of data regarding the incidence of the phenomenon. Insurance coverage may be offered by national organisations but more generally by private insurance companies. Profits are not meant to be made by the insured from insurances, so that the hazard should not be capable of being wilfully created. There should be a need on the part of the insurer to take out insurance to ward off loss of income and he should have sufficient finances to meet premium payments. Where it is in the national interests to insure against loss, government agencies may assist to finance schemes. The Israeli Government supports such a scheme, financing up to a half of a national crop insurance facility.

Insurance is really the substitution of a small cost which features as part of the firm's production costs for the possibility of a larger uncertain loss. This definition might also embrace the use of crop protection chemicals. Although the incidence of a pathogen may fluctuate, the need for using the chemicals is often more regular than it would be in our true insurance situation.

After taking out an insurance policy the grower will always wonder what his income would have been if he had saved his outlay on insurance. A grower with glasshouses or crops shattered by hail may ruefully lament not having entered into a hail insurance. Many years may elapse without recourse to claim from the insurance company and in horticulture with its characteristic longer term investments in glasshouses and orchards the temptation not to take out or to renew a policy may be strong.

Without too much concern for the 'small print', the different types of insurance available are considered below

a. *Crop insurance.* Hail, flood, drought or frost all constitute uncertainties for the grower. Hail is the commonest of these for which insurance schemes have been available in France and Germany since the late eighteenth century. In most countries where these schemes operate the insurance is written by joint stock companies and mutual insurance societies together with friendly or cooperative insurance societies. In the UK hail insurance is available from an insurance company owned by the National Farmers' Union and other schemes from specific insurance companies. The county of Suffolk experiences more hail than any other county in the UK and premiums are therefore higher there. Other counties can be grouped and premiums charged according to their relative susceptibilities. Other insurance schemes are available in the UK for spray damage to crops, storms, fire, lightning and aircraft. Sometimes a policy may cover two or more hazards and may specify one or more crops. In a few countries all-risk schemes have been evolved, a comparatively recent development often associated with Governmental policy towards agriculture.

b. *Property insurance.* Fire insurance is the oldest and most extensive form of insurance in this category. It is practised in most countries of the world especially the developed ones. Cover always includes ignition by lightning and may include the explosion of boilers and even damage by aircraft and other aerial devices. Generally it does not include civil riot, enemy hostilities, earthquakes or spontaneous fermentation. Insurance on farm property may include buildings, household and personal effects, machinery and animals, live or dead.

Agricultural vehicles are increasingly dangerous to operate and costly to buy or replace. Insurance is therefore necessary to cover loss or damage and death or injury to operators or other persons resulting from their use. Insurances of this type are required by law in most countries for road vehicle users.

c. *Personal and liability insurance.* These provide protection against personal accident or sudden illness and are distinct from national and social insurance schemes. The liability of employers extends to employees and to members of the general public should they be injured as a result of farming activities.

For the employer himself, the policy might include a life cover to make provision for his dependants. This insures against 'living too long' or of dying too early.

2. *Choosing relatively safer crops*

The yields and prices of some crops vary greatly from year to year and even over shorter periods. Others by contrast are more stable over the years and the grower desiring stability of income can choose to grow these more regular yielding crops or cultivars if he is aware which they are. Table 4.2 contains some examples of variability in crop yields. These indices have been calculated from twenty years' data and can be regarded as a 'league table' of variability under conditions prevailing in England and Wales. The protected environment of the glasshouse ensures the minimum variability, whilst the inflexible long term investment in tree crops exposed to all weathers, makes for greatest variation. It is particularly interesting that glasshouse tomatoes and outdoor tomatoes should be top and bottom respectively in the vegetable part of the table. Outdoor tomato growing has declined almost to extinction largely because of uncertainty and even frequent total failure in some areas.

Table 4.2

Yield variability indices for selected crops in England and Wales

	Variability Index † *(Based on 20 years' data)*
(a) *Vegetable Crops*	*per cent*
Tomatoes (heated glass)	2‡
Cucumbers (glass)	7
Beetroot	8
Onions (dry bulb)	10
Leek	10
Brussels sprouts	13
Cabbage (spring)	14
Asparagus	19
Cauliflower (winter)	22
Tomatoes (in open)	41 §
(b) *Fruit Crops*	
Apples (dessert)	12
Strawberries	13
Blackcurrants	24
Apples (culinary)	25
Cherries	50

† Variability index = residual standard error divided by the 1962-63 – 1967-68 mean.

‡ Based on ten years' data.

§ Based on seventeen years' data.

Stanning, P. J. (1971). 'Project Report for the Degree in Horticulture' (Unpublished). University of Bath (carried out under the author's supervision).

3. *Specialisation or diversification*

The prevalence of variability that we have identified in the last section means that if only one crop is grown (mono cropping) then incomes will fluctuate accordingly. This is the ultimate in specialisation. Inevitably it makes for the highest uncertainty but might also be the most profitable choice measured over several years. Growers commonly grow several crops as a precaution against this uncertainty. They avoid 'putting all their eggs in one basket' and diversify. Indeed, some diversification up to say, three enterprises, may lead to higher profits. There may well be complementary features between the enterprises that make them a particularly viable combination. Examples are that a large amount of labour is required in the summer for hops and in winter for apples and pears. The different crops grown on the same holding in balanced proportions can utilise labour very efficiently.

The general basis for diversification is that if one crop does badly at the market another will probably be successful. It is unlikely that all crops will realise low prices except in times of general depression when diversification gives protection. The grower may choose one relatively stable enterprise to ensure some income and then speculate on other more uncertain ones. This may conflict with rotation on the holdings for rotations are closely associated with diversification.

By specialising the grower can put all his resources and expertise into producing a large volume of produce. This may be desirable if a processing outlet is being supplied. Furthermore, it may be easier for the grower to keep abreast of technical developments in this narrower field. This may be reinforced by assistance from the processors, themselves concerned with maintaining the quality of raw material arriving at their factories. The production of large quantities of standardised produce implies specialisation and therefore more uncertainty, so it is not surprising that specialisation has been accompanied by contractual agreements between growers and large scale buyers.

4. *Contract growing*

We discussed contracts in chapter 3. By entering into a contract for the sale of his produce in the future the grower can eliminate a great deal of uncertainty. He does not always have this opportunity and the contract may only favour him according to the relative bargaining strengths of the parties negotiating it. In any case, the grower can only forward contract his prices; costs may still fluctuate but will most likely increase.

5. *Choosing flexibility in production*

By avoiding rigid, inflexible methods of production, further precautions against uncertainty are made. The purchase of an expensive harvesting machine capable of harvesting only one crop in very exacting conditions might represent inflexibility. A harvester of more universal application might offer greater flexibility. Some further examples might include the use of dual purpose cultivars, the maintenance of a pool of readily available resources for emergencies or the deliberate choice of a short term rather than a longer term crop.

It is overheads or fixed costs (see chapter 6) that weigh heaviest in uncertain situations. These costs must be met by the grower whatever enterprise he chooses. Once he has purchased machines and buildings they will depreciate, permanent labour will require payment, rent and office expenses must be paid. In times when prices are lower than expected these costs become difficult to meet. As a general rule, the grower with a small amount of capital and large debts to meet would be advised to opt for the more flexible growing system. In contrast the grower with ample capital and small debts can cope better with inflexibility.

Day to day business plans may be kept flexible so that as the need arises new plans can be formulated and direction changed. An inflexible fixed plan of action may not be able to take account of periods of prolonged drought or rain, of machinery breakdowns or even sudden illness on the part of operators or growers.

We have shown that uncertainty is the dominant feature affecting the grower's decisions and actions in combining resources or inputs. In the next chapter we consider the three broad categories of resources used in production.

CHAPTER 5

RESOURCES

All the inputs used in the growing of a crop can be classified under three categories – capital, land and labour. These are the resources or factors of production. A price is paid for the use of each of them determined by a supply and demand interaction but all are scarce in supply. The prices of capital, land and labour are called interest, rent and wages, respectively. A grower invests his capital in an area of land and further introduces his own or borrowed capital in seeds, fertilisers and tools. He makes these productive by the application of his own or purchased labour. All are involved in production but we consider them separately here for convenience.

5.1 CAPITAL

Use of capital

The amount of capital accumulated in a business is represented by the land, buildings, equipment and growing crops. Financial values can be attributed to these and appear in the 'balance sheet' account of the firm. Two types of capital are commonly identified: *real* or *fixed* capital which includes land, buildings and equipment and *working* or *circulating* capital mainly comprised of labour and raw materials used in the productive process.

The grower needs capital in his business for three purposes. Firstly he may wish to expand the business. This is the most common purpose and may involve the purchase of a new holding, an additional field, glasshouse or machine. He increases the capital stock of his business. Secondly he may be concerned with a replacement investment to maintain his production. A glasshouse by obsolescence or an orchard by senescence may be drawing to the end of a useful life. Thirdly the entire industry may be under pressure to innovate. Prices of glasshouse vegetable crops may be so low that growers are forced to consider and try alternative crops. They may turn to flower production or even develop completely new crops.

The process of injecting further capital into the business is a fresh additional investment not to be confused with the narrower stock exchange concept of investment in stocks and shares. An investment is the spending of money in the present with the aim of securing a stream of earnings in the future. For example a grower may invest today by planting an apple orchard with a view to receiving a succession of crops and financial returns for many years in the future. He can obtain capital for this venture by transferring some of his own profits or by borrowing from friends, banks or other credit organisations. He pays a price for the use of the capital. To understand investment better it is important to consider how the price is determined, the sources of capital in society, and how it is rationed.

The source of capital

If capital were plentifully available it would be borrowed much more freely. The supply of this scarce commodity is dependent upon several factors. So that capital is available for borrowing someone must save or give up consumption. If we spend our entire income as we receive it our

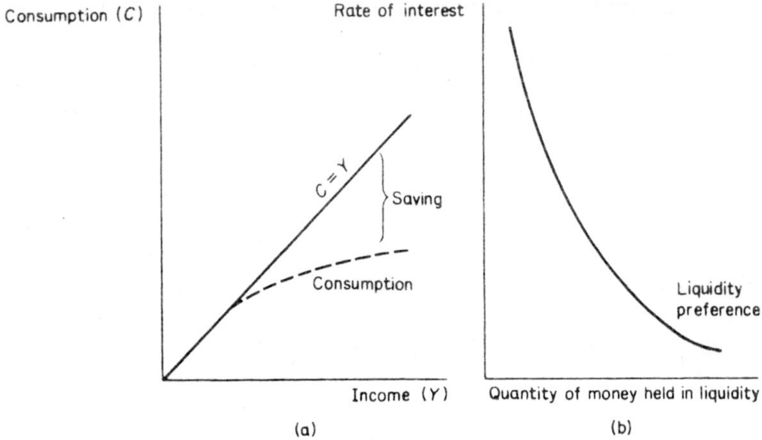

Figure 5.1(a) and (b)

consumption is equal to our income. However if we set some of this income aside for use in the future this constitutes a saving. Figure 5.1(a) illustrates this diagramatically, the straight line representing increase in consumption as income increases; it is often referred to as the *marginal propensity to consume* (MPC). In this case all income is consumed and MPC = 1. The broken line follows the path of a consumer who saves an increasing part of his income as it increases. Generally as incomes increase in a community consumption will expand but at a slower rate.

That money cannot be loaned without someone parting with it for a period of time may seem obvious. It is an important point, not always appreciated and is the main explanation of the restriction in supply of capital. If capital is held either as cash or in a form that can be easily transformed into cash then the particular individual or organisation is said to have a high *liquidity*. If money is loaned liquidity is given up. This in itself leaves the lender at the mercy of changes in economic circumstances and he may require payment for the inconvenience. In addition he hopes for a favourable outcome in the form of high dividends and increasing values. Each lender has a particular *liquidity preference* which changes according to the potential earning power or prevailing interest rate. If rates are high then he may prefer to lend in response to the offer of a high return but as rates fall the benefits from lending decline and greater preference may be given to liquidity.

The quantity of money available overall may be controlled by the central government which will determine prevailing interest rates. The liquidity preference may vary with change in interest rate as in figure 5.1(b).

The rate of interest

The demand for the use of capital by borrowers depends upon the prospects they envisage for its use. If indications are that prices for a particular crop will be high in the future then growers will be tempted to invest. The reverse is true in times of price depression. The interaction of supply and demand results in a price, or interest rate, and might be represented as shown in figure 5.2.

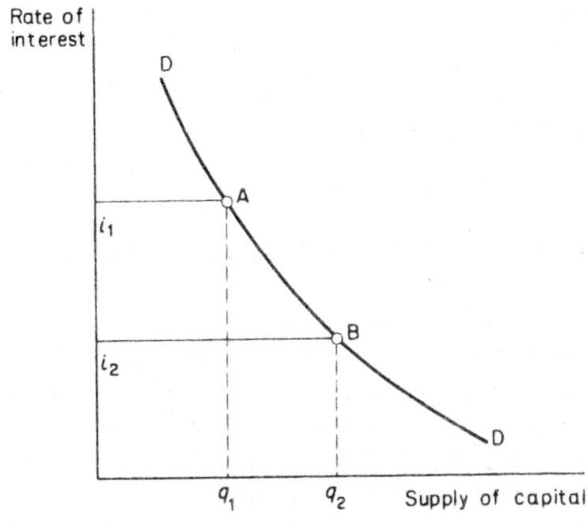

Figure 5.2

When a quantity of capital q_1 is available the interest rate is high at i_1 but as the quantity available is increased to q_2 the interest rate falls to i_2, the demand curve for capital DD remaining static.

There is no such thing as *the* interest rate; the mechanism is dynamic and continually changing. Rates move up or down according to the general buoyancy of the economy.

Investment in horticulture

The specific cost of a new investment and the performance that can be expected from it may be generally available. There are however other features of the investment that should be considered but are often overlooked. In considering any future investment four basic interdependent questions ought to be posed by the manager.

1. Is it worth making the investment at all?
2. What is the optimal economic life of the new investment?
3. When is the best time to make the new investment?
4. What is the benefit of a replacement investment compared with maintaining the old?

These are simple questions but in practical horticulture they are commonly difficult or even impossible to answer satisfactorily. Let us consider two situations to try to understand why this should be.

Table 5.1

Estimates of the amount of fixed capital used

	Fixed Capital†	
	per employee £	*per acre* £
Manufacturing industry	2100	–
Chemical industry	8000	–
Agriculture	6500	300
Fernhurst orchard	4500	720

† Land, buildings, plant and machinery

Cobham, R. O. *et al* (1966). *Agriculture* 73, 8, 361.

In the first, a manufacturing company installs a machine to produce a steel bolt. The cost is easily identified and the machine can be installed and production commence virtually overnight. Output from the machine is predictable providing that it is maintained and supplied with steel. To the end of its economic life, also predictable, it will go on producing bolts at the same rate. Compare this with an investment by a fruit grower who decides to plant an orchard of plums. He chooses his trees from over a hundred cultivars often not well tested or proven in performance. They are planted outdoors exposed to all weathers and may take as long as ten years to crop. In each year thereafter the total crop will depend upon weather together with the grower's skill in husbandry. Crop yields and value may vary considerably from year to year and cash be received for a relatively short period of the year when the crop is marketed. The natural life of the plum tree may be well over a hundred years but the economic life which is dictated by several factors may be much shorter, perhaps 20 or as long as 70 years. The orchard may fail or prematurely age due to soil or weather or the cultivar may become obsolescent in the market. These are extreme investment situations. However, if the reader applies the four simple questions posed at the beginning of this section to these examples, he will be left in no doubt of the uncertainty facing the horticultural investor.

As an industry concerned with intensive crops horticulture uses large amounts of capital and it is surprising to discover that few analyses have been made of the extent and efficiency of use of capital in the industry. More information is available for agriculture as a whole.

Table 5.1 is an attempt by some economists employed by Imperial Chemical Industries Limited to compare the relative employment of capital in different industries in the UK. The table includes the ICI dessert apple orchard at Fernhurst which serves to illustrate the relatively high capital intensity of horticulture. In terms of investment per employee it is not as high as the chemical industry but in terms of unit area is much higher than agriculture in general.

Investment and technology

The importance of agriculture as a source of capital, particularly in the transitional stage was stressed in chapter one. As growth proceeds capital progressively becomes more prevalent in relation to labour in the economy. The replacement of labour with capital is still a prime aim of policy even of developed countries. Failure to inject sufficient capital leads

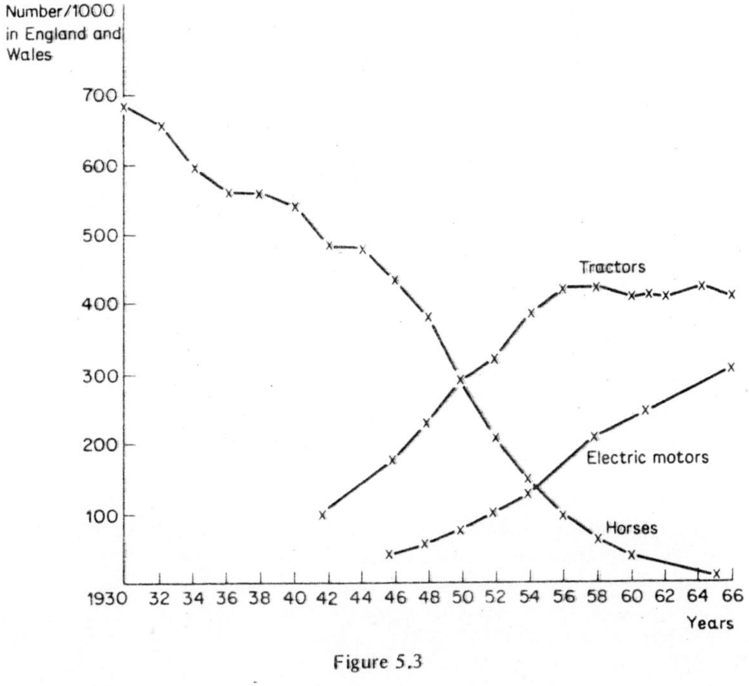

Figure 5.3

to unbalanced economic growth. Horticulture may have both high investment and labour usage particularly in protected cropping and in this respect is of particular interest as an industry.

Investment in the agricultural sector itself is an essential feature of technological advance. New methods of production may bring new habits and a new set of requirements on the part of the population usually implying further capital spending. In this way it is an accelerating process. The growth of particular forms of technology and then later decline and supersession by new developments take a characteristic form. An illustration of this is the development of the farm tractor leading to the decline of the horse. Figure 5.3 depicts this situation for the UK over the past 40 years. The characteristic flattened 'S' shape of the curve for the growth of the new technology is plain in relation to the decline of the horse. Electric motors have been included as an even newer technological advance but the curve has not yet reached its maximum.

The time value of money

Investment involves spending and receiving benefits over a period of time. If you are offered a bank note to take now or in a year's time you would be wise to take it now because you can invest the money so that in a year it will be worth more to you. Furthermore, not only are you wasting its earning power if you do not use it but you have the responsibility for its safety. We say that money has time value and to make comparisons between investments for the future, often of different duration, we need to be able to relate the value of future anticipated benefits to current values. We use a process called *discounting* in investment appraisal to do this. Similarly, if we wish to estimate the future value of a current investment the reverse process is carried out and is called *compounding.*

Discounting, compounding and valuation

The present value V of a promised income or return R is found by substituting relevant figures in the equation $V = R / [(1 + i)^n]$ where i is the interest rate received over a period of n years. For example if £100 is promised at 5 per cent interest in 10 years then the present value is, $100/[(1 + 0.05)^{10}] = £61.3$. The £100 in the future has been discounted by the appropriate discount rate to a present day value.

In the production of tree crops several years may elapse before a return is received. Investment may be spread over the interim period and the grower may wish to know the extent or value of his investment by the time the first harvest is made. For example if a grower invests £100 now in an orchard and the alternative investments open to him could yield say 5 per cent interest, then in a year's time the investment will be worth £105. In the second year the cost will be compounded from the first year so that by the end of the second year the investment is worth £105 plus 5 per cent interest or £110.25.

A generalised formula for compounding can be written as $C = E(1 + i)^n$ where C is the future value of some current cost E considered over n years at interest rate i. Some horticultural investments are typically made over several years as with our orchard example. If E_1 $E_2 ... En$ are expenditures at different times, the equation can be slightly modified, thus

$$C = E_1 (1 + i)^n + E_2 (1 + i)^{n-1} .... En (1 + i)$$

Appreciation and depreciation

The calculations of the last section showed how present values of an investment may increase or decrease with time. These are called *appreciation* and *depreciation* respectively. Depreciation is the most commonly encountered and refers to the loss of value of an asset in relation to its use. Over a period of time obsolescence occurs regardless of the amount of use and is therefore a fixed and inescapable cost which has an important effect upon personal and business incomes. The incidence of taxation of incomes in some countries varies according to allowances made in accounting for depreciation.

Depreciation is most commonly calculated by discounting the investment at either a fixed amount each year or by a percentage. If we invest £100 today in a machine we may choose to depreciate or 'write off' say £10 per annum so that it is apparently valueless after 10 years. If we use the fixed percentage method, say 10 per cent, then after 10 years the machine is still apparently worth £33.56. In this second situation, depreciation is rapid at first but declining in rate with time and never reaches a zero value.

The most usual investment situation is where a single outlay is made and depreciation occurs thereafter. Perennial crops are exceptional in that there is an appreciating phase before depreciation and there is a series of annual yields. Forest crops have a terminal yield.

Investment appraisal

To assess the likely outcome of an investment project and to compare projects the manager requires accurately appraised investments. All too often the methods used in practice obscure the true outcome even though detailed considerations may have been given to likely costs and returns. Failure to take the timing of expenditures and receipts, taxation and the availability of investment grants into consideration is common.

One method still used in horticulture is the very traditional 'return on capital', which involves expressing the anticipated extra profit as a percentage of the capital cost. Suppose a glasshouse grower erects a glasshouse costing £10 000 and spends a further £2000 of working capital in the first year. If £2000 are expected at the end of the first year as profit then return on total investment might be calculated as 2000 x 100/12 000 = 16.6 per cent at the end of the first year. The calculation is simple but masks some important features. The investment will undoubtedly last for longer than a single year and no account has been taken of depreciation of the initial investment. The longer the life expectancy of the project the lower will be the annual cost of depreciation. Working capital and fixed capital have been treated alike although the former is recoverable and the latter is not. No account is taken of the timing of the investment and the receipt of payments. If this method is used it is imperative that the origin and limitations of the figures used in the calculation be defined.

A more modern and improved technique is the method known as discounted cash flow (DCF). As the name implies this technique is concerned with the net cash flows of a project – the difference between cash receipts and cash expenditure in any year. These are characteristically negative at the start of the project when money has been paid out and positive when money accrues to the firm.

DCF makes use of the discounting principle and by specifying a particular discount rate we can calculate how much a promised cash flow is worth now. Discount factors are used for this purpose and they represent the present value of £1 received *n* years hence at a particular discount rate. These are laborious to calculate and are fortunately available as tables. The discount factors are used to discount each cash flow as illustrated in the very elementary example calculations in table 5.2.

Table 5.2

Investment appraisal. Discounted Cash Flow (DCF)

Year	*Annual cash flow expected*	*Preferred rate at 15 per cent*		*Preferred rate at 20 per cent*	
		Discount factor	*DCF*	*Discount factor*	*DCF*
	£		£		£
0	−14000	1.000	−14000	1.000	−14000
1	+ 4000	0.870	+ 3480	0.833	+ 3320
2	+ 4000	0.756	+ 3040	0.694	+ 2760
3	+ 4000	0.658	+ 2640	0.579	+ 2320
4	+ 4000	0.572	+ 2280	0.482	+ 1920
5	+ 4000	0.497	+ 2000	0.402	+ 1600
6	+ 4000	0.432	+ 2580	0.335	+ 1980
	Net Present Value (NPV) =		+ 2020	=	− 100

In this table an initial investment of £14 000 has been made and appears as a negative sum. The preferred earning rates of capital have been set at 15 and 20 per cent respectively and for each value the appropriate discount factors abstracted from tables. Each cash flow has then been discounted by multiplying the cash flow by the discount factor. The totals of the DCF columns give the net present values (NPV).

From this analysis we can draw some useful conclusions about the project. At 15 per cent the NPV shows a surplus after the deduction of the £14 000 initial investment. Put another way, money could be borrowed for this project at 15 per cent interest and there would still be £2020 surplus at the end. At 20 per cent a deficit occurs. Ideally, we are seeking an interest rate which yields a zero solution because this will represent the true earnings rate of the project. This is clearly between 15 and 20 per cent in our example and very close to 20 per cent. This appraisal has been far more comprehensive than the 'rate of return' calculation discussed earlier. It has taken account of the timing of the cash flows and in addition has given us information about borrowing rates that can be afforded.

To choose between alternative investments the pure technique is applied to each one and a direct comparison is made of the NPVs. The highest NPV at a given interest rate will yield most profit though it should be noted that this is not necessarily the project yielding the highest return on capital. Projects ranked by NPV may cost different amounts to start. Further improvement in the appraisal can be introduced into the determination of expected cash flows. Allowance can be made for extremely uncertain situations, such as very variable prices, inflation, taxation, and the availability of investment grants. These adjustments are not discussed here but are described in most modern books on investment appraisal.

Very long term investment

The DCF technique is very appropriate for investments of the depreciating type. With orchards the time span of the investment might be 50 years or more. Uncertainty of the future considered over this length of time will be high. A grower making an investment for the first time may find the NPV criteria worth while considering. The grower who already has established orchards may be loath to fix a time period over which the investment must be paid off. If a time allowance is made for the likely uncertainty together with the alternative uses of capital then the interest rate used in discounting may rule out the planting of an orchard which in the long term may yield very satisfactory returns. Caution is therefore needed in applying appraisal techniques to this type of situation. The current trend towards smaller shorter-lived types of trees increases the capital invested because more trees are used. These are dependent upon an early return on capital, increasing the uncertainty of the project. Moreover the life-spans of these new forms of investment have not yet been completed or proven in reality.

5.2 LAND

Capital and land are closely related. Land accounts for the greatest part of capital assets in almost all countries and the agricultural industries are by far the dominant users of land. At the end of the nineteenth century approximately 90 per cent of the UK land surface was in some form of agriculture. It was still 81 per cent in 1965 even after large losses to urbanisation.

Price determination

Like other factors of production land is scarce. For any particular country the total area is generally fixed and only rarely is the entire area suitable for farming. Some reclamation of land from the sea and from open cast extractive workings has occurred but is very small in relation to the total area. The land area available in relation to population is therefore very different from country to country. Table 5.3 illustrates this for some countries having a significant horticultural area. It does not take account of the differing quality of land and quantities available in the countries.

Table 5.3

Land† use intensity for some countries prominent in horticultural production

	Population density (hectares available per person)
Taiwan	0.26
Netherlands	0.26
Belgium	0.32
Japan	0.36
UK	0.43
Western Germany	0.43
Italy	0.55
Israel	0.73
France	1.09
Romania	1.20
Cyprus	1.48
USA	4.57
Kenya	5.57
New Zealand	9.66
Australia	64.16

† *total land area excluding major rivers and lakes.*

Adapted from FAO (1969). *Production Yearbook*, 23.

Different types of land are clearly substitutes for each other to varying extents but there is no substitute for land so supply is highly inelastic. Marginal land may be brought into cultivation or fallowed according to the general prosperity of agriculture and there will be interchanges between types of agriculture as the relative prosperities of particular enterprises fluctuate. Vegetable crops may well be grown to an increasing extent by cereal growers as the relative price that they receive for cereals falls. A period of prosperity for horticultural crops may lead to an expansion of the acreage even on to land not entirely suitable for the production of these crops. Growers here are the first producers to suffer on the resumption of lower prices or in a general depression.

Demand for agricultural land depends upon the general prosperity of farming and the number of prospective farmers or growers. The quality of land in relation to particular uses also affects demand. It is pressures from alternative uses to agriculture that become dominant in countries of high population density. Urban development in the form of housing, offices, airports, roads and reservoirs is particularly important in this respect. Higher incomes and incidence of car ownership have led to greater mobility which together with increased leisure time has led to a new pressure, that of recreation.

Price is determined by the interaction of supply and demand and as supply is virtually fixed it is demand that mainly determines price. The market is not as simple as it appears because land transactions are not confined solely to the land itself; they include buildings, trees, mineral deposits and water resources. They are really dealings in bundles of rights over the use of land. Price differentials for land may be due entirely to competing interests in the use of land. Land near an urban centre with permission granted by a planning authority for development may take on a greatly increased value due to demand from builders and developers and may be worth many times the agricultural value.

It was the nineteenth century economist Ricardo who drew attention to the fact that rents of lands arise essentially out of differences between them and that high rents reflect the 'niggardliness of nature' in not providing sufficient land of adequate quality for everyone. As demand is influenced by the prosperity of agriculture we would expect land of the best quality and cropping potential to fetch the highest prices because it will be used for intensive production.

Table 5.4

Agricultural land value by type of holding (England and Wales) 1962-64

Type of holding	*Mean price per acre £*
Bare land	107 ± 5
Stock farm	160 ± 15
Dairy	174 ± 8
Arable	189 ± 8
Mixed or other	198 ± 5
Smallholding	233 ± 15
Fruit, Hops, Market Garden	238 ± 19

Clery, P. A. and Wood, P. D. P. (1965). 'Agricultural Land Values 1962-64'. *Journal of Agricultural Economics*, 16, 4, 567.

Table 5.4 shows some difference in the prices of holdings grouped by type. The size of holding is also influential, prices per unit area declining as size increases. Lastly, land with vacant possession that can be taken over immediately will fetch higher prices than land which is tenanted and therefore encumbered. This is illustrated in table 5.5 which also reflects rental payments in relation to the net farm income.

Table 5.5

Land market indicators (England and Wales)

Year	*Land value £ per acre*		*Total net farm income £M*	*Average Gross rents £ per acre*
	Vacant Possession	*Tenanted*		
1937-39	25	23	56	1.28
1949-51	82	46	331	1.68
1958	85	53	378	1.92
1960	123	61	397	2.35
1965	235	150	464	3.55
1966	242	154	492	3.80
1967	258	135	510	–

Peters, G. H. (1968). 'What Price Land?' *Br. Farmer*, April 6, 64.

Land tenure

Land prices have been discussed so far in relation to change of ownership of land but land is also used or rented at a price without change of title deed. Classical economists referred to rent as being any surplus that accrues to a factor of production: it did not only apply to land. For our purposes we shall consider rent as a payment for the use of land and other associated facilities for a defined period of time. Economists talk of *economic* and *contractual* rent. Economic rent is often defined as the residual income to land after the deduction of payments for all other inputs from the gross product. It is a theoretical level determined largely by supply and demand. Contractual rent is also determined in this way but it represents the payment that is actually made and includes the use of buildings and equipment. Payment is usually fixed in financial terms but examples of other types include tithes, payment by services provided and share-cropping, where the landlord takes a proportion of the crops produced from the land.

The terms of occupancy of land are known as the system of *land tenure.* This differs from country to country. In primitive societies very few individually owned holdings are found. In the UK the feudal system 'broke down' and in the eighteenth century enclosures of land created estates of differing sizes each comprising several holdings. A unique system of landlord/tenant agreements grew up. The larger landowners rented holdings to farmers who paid annual rents for the use of the properties. In 1900 80 per cent of holdings under 5 acres were tenanted but by 1961 this had fallen to 45 per cent, owner occupancy becoming more prominent.

The method of tenure can be critical in economic development. It can impede progress particularly where custom requires the division of holdings between the children of a family on the death of the father and farm sizes get progressively smaller. Another example occurs where no compensation is made when tenants who have made capital improvements have to give up their holdings. This discourages technological advance and may be particularly problematic in horticulture with characteristic longer term investments in orchards and protected cropping. There is little incentive for a grower renting a holding to invest in this type of crop. Successful schemes have been adopted in some countries not only to combat this last problem but also to provide state owned holdings leased to tenants. These may serve an important function by affording easier access for new entrants to the industry.

Owner-occupancy confers distinct advantages in that improvements made to the holding benefit only the owner. Security for himself and his next of kin together with freedom to control may be all important goals for the owner even though they may have little financial value. A real disadvantage is the need for a large amount of capital. Even after purchase of a holding with mortgaged ownership, which is most likely in western European agriculture, it may be difficult to find the working capital still needed.

In terms of business success the advantages of owner-occupancy or tenancy are not always clear. Different managerial ability may be needed for the two systems of tenure. Some landlords may select only men of exceptional ability to lease their holdings, and some men of lower ability may only be able to farm by purchasing their own holding.

Taxation

Horticulture commonly evolves in close proximity to urban centres. Increasing urbanisation and the expansion of these centres can lead to strong pressures on these producers to 'sell up' and move away. Prices offered for land can be very high so that a substantial sum of capital can be raised from the sale of the holding. Many countries operate some form of taxation on capital gain that may arise from the sale of land or estate or by inheritance. Special concessions to prevent the loss of capital from agriculture may be given to farmers which may further influence land prices.

5.3 LABOUR

Labour is necessary to make capital and land productive. When we talk of labour we are generally referring to the hired workers employed by the owner of the holding. In primitive societies and in some parts of developed countries these may be non existent, family organised holdings being dominant.

Horticultural work

Horticultural work has some different characteristics from other industries. Firstly there are prominent seasonal peaks of labour requirement, usually at sowing or planting and harvest. A labour force may be employed to cope with these peaks but be under utilised for parts of the year. Horticultural firms generally employ more labour than agricultural ones which may accentuate this feature. Part time farming is one solution to this problem, the employer taking a public service job for a proportion of the day or year. He may even take a regular job, working his holding in his spare time and employing temporary labour, supervised by his wife or other family member, at peak times. Small holdings growing strawberries are often organised on this basis in the UK. Another solution is to diversify and grow more crops which may extend the sowing and harvesting seasons so that the labour force is more regularly employed throughout the year. Most vegetable growers produce more than one crop to avoid price uncertainty and to achieve a balanced use of labour. Bush fruit growers in the UK commonly claim that they grow several different crops "to keep the pickers on the farm". This is to prevent the loss of scarce casual labour to competitive farmers. The employment of many casual labourers is a common feature of horticulture and is another way of coping with the work peaks. In some countries bands of migrant workers make themselves available for seasonal work. Housewives and students are important as casual workers. Contractors are used by growers who may not own their own machinery so there is a further advantage through less need for capital investment. A serious and often inhibiting drawback is that several growers may all require the service on the same day so that where highly perishable crops are concerned some loss is inevitable.

The second major difference between horticulture and other industries is in the size and number of production units. Horticulture is predominantly comprised of small firms employing up to two or three workers. Only in the developed countries have firms evolved with large payrolls.

The increasing sophistication of horticulture leads to the need for training and experience amongst 'key' workers. The worker may find himself in sole charge over a weekend of a glasshouse holding with fully automatic heating, ventilation, irrigation and nutrition controlled by highly sophisticated electronic apparatus.

Wage determination

Wages are the price paid by the manager for the use of labour. Formally defined, a wage is a sum of money paid under contract by an employer to a worker in exchange for services rendered. The contract is generally very informal and wage determination in agriculture probably approximates to a pure market.

In most situations, actual wage rates paid are arrived at by negotiation and are determined by the number of workers seeking employment in agriculture and the demand for their services. Demand is a function of the general prosperity of agriculture, the need for workers' services and the ability to pay for them. Trade unions are not strongly developed in agriculture but in many industries trade unions act as supply control agencies in the labour market. By operating as the sole negotiator on behalf of workers and by threats to withold labour, minimum wage rates are determined. The National Union of Agricultural and Allied Workers (NUAAW) although one of the largest unions in the UK, has probably not more than 70 per cent of its members actually working in agriculture. The remainder work in industries auxiliary to agriculture and in other occupations. No figure is available of the proportion of horticultural workers in the Union in relation to the total. The rural widely dispersed nature of the employment makes it difficult for the Union to recruit and communicate with members and collect subscriptions. Despite these problems the NUAAW is the chief negotiator on the Agricultural Wages Board for minimum wage rates in the UK. The employers are also represented by a single Union, the National Farmers' Union (NFU).

The Board comprises eight employers, eight employees and five independent members and meets when agreement is reached to open negotiations. The workers make their claim for a wage increase, usually based on the disparity of agricultural wages with other industries and the increase in cost of living. They further claim the ability of the employers to pay higher wages because of increased productivity and maintain the need to halt the decline of the labour force by greater financial incentive to stay on the land. The employers argue that continually rising costs with no increase in real prices for their produce leaves them unable to afford higher wages. They stress the additional benefits workers receive as agricultural employees such as free or low cost housing, produce and fuel. Disagreement is inevitable so that the independent members determine the wage settlement.

It is only a statutory minimum wage and weekly hours of work that are laid down by the board. No wage structure exists although at the time of writing one is proposed in the near future. Actual earnings in agriculture usually exceed the statutory minimum, the excess being made up of bonus earnings, premium payments and overtime. For the year ending March 1970 workers in horticulture in the UK actually earned less than the average for all hired men in agriculture and horticulture by £0.83 but average time worked weekly was 2.3 hours less. The data presented in table 5.6 detail the actual cost of labour to employers in certain European countries recently for three categories of workers according to training and sex. There are computational difficulties in making international comparisons of this type. In some countries payment for training is not commonplace and equal pay for men and women is not usual.

Table 5.6

Actual cost of labour† *per working-hour in some countries in Western Europe (Swiss Francs) 1967*

Country	*Trained horticultural worker*	*Untrained older horticultural worker*	*Female horticultural worker*
Sweden	8.10	7.44	6.75
Switzerland	7.13	6.28	6.28
Norway	6.08	5.90	5.08
Denmark	6.00	5.92	4.92
Netherlands	5.51	5.57	5.62
Western Germany	5.14	5.05	4.14
France	4.03	3.91	3.46
Great Britain	4.04	4.07	2.83
Belgium	5.71	5.28	5.38

† Hourly wage plus pay for vacation and general holidays for the workers and social charges paid by employers without cost of sickness and lost hours and without irregular gratuities.

The International Association of Horticultural Producers (AIPH).

Table 5.7

Agricultural workers in Great Britain 1801-1965

	Number employed (millions)	*Proportion of occupied population %*
1801	1.7	35.9
1851	2.1	21.7
1901	1.5	8.7
1951	1.1	4.7
1965	0.9	3.4

The declining labour force

In chapter one we saw the characteristic decline in importance of agriculture as an occupation during economic growth. Table 5.7 illustrates this trend for the UK over the past century and a half.

The horticultural labour force has also declined but at a slower rate than in agriculture generally. Horticulture is characteristically an intensive user of labour. Cropping under glass may involve 10 to 12 men per hectare and the fruit grower might need a man per 8 hectares. A cereal grower by contrast might need only one man on 60 hectares. Horticulture had 17 per cent of the regular whole-time labour force employed on all agricultural holdings in the UK in 1967.

Economists have identified features that influence the movement of labour from agriculture. They talk of 'push' and 'pull' effects. The former are internal features encouraging movement away and the latter are effects enticing outward migration. It is the hired, regular, whole time workers in the UK who form the greatest proportion of migrants. In the USA by contrast it is family workers who migrate most strongly.

The dominant factor influencing labour migration is relatively attractive wage rates offered in alternative employment. As the disparity between wages in agriculture and elsewhere increases so migration increases. As the general employment level of an economy rises so migration rates are also encouraged. In the USA in the depression years of the 1920s labour actually flowed back into agriculture. Clearly awareness of alternative opportunities by would-be migrants and the distances to be travelled are important factors. The close proximity of alternative employment makes migration easier.

Young workers, particularly after receipt of higher education are most likely to migrate. The more educated the worker becomes the more aware he is of alternative wage rates and the greater his aspirations. Wage rates in horticulture may be lower than in other industries for young unskilled workers and this encourages migration.

Social features may also be important. The type of quality of housing provided or available in rural areas may leave much to be desired. The remoteness of production areas may also raise problems of travelling for shopping and education of the children. The work itself although having the obvious attraction of the outdoor life can at times be arduous because of long hours and inclement weather.

The employer gains increasing incentive to substitute capital for labour as the cost of labour increases. He does this by investment in machinery. If capital is expensive in relation to labour then this will be retarded. Labour accounts for 30-50 per cent of total production costs so he may seek to improve the productivity of his existing labour force by resorting to management techniques such as work study.

As labour migrates from agriculture greater productivity is achieved, particularly if the output continues to increase. In the UK labour productivity of agriculture has for fifteen years been one of the highest in all industries. During the period 1948-65 real net output in agriculture increased at approximately 2.7 per cent per annum but the labour force declined by 2.5 per cent per annum. G.J. Tyler of the University of Oxford estimates that during this time the real capital stock increased by only 0.7 per cent per annum. This represents phenomenal growth in productivity.

Workers are not all equal in ability and work potential and are therefore not perfect substitutes for each other. The casual fruit picker may not have the same skills as the man who is in charge of the sophisticated glasshouse environment controls. He may of course be equally intelligent and merely lack training and experience. Managerial and supervisory ability are less easily taught than technical skill.

Supervisory staff foremen and chargehands must emerge from the general work force of the industry as the need arises for their services. Managers of the right calibre and training must be attracted to the industry to control the business, often on behalf of absent owners.

CHAPTER 6

ORGANISING PRODUCTION

To use the resources discussed in the last chapter the manager incurs a cost, the price that he pays for them. Cost varies with the volume of production over a particular time period.

6.1 THE BEHAVIOUR OF COSTS

Opportunity costs

The grower is commonly concerned to know the cost of following a line of action. To estimate this satisfactorily he really needs to look at the alternatives which have to be sacrificed if the particular line is followed. He may for example buy a particular brand of fertiliser (A) at say £20 per ton. If another brand (B) at the same price would give him £5 extra in returns, then he has really incurred an extra cost of £5 by not choosing brand (B). We call this cost the *opportunity cost.* Another example might be that of a grower using his own money to buy a new implement. He might suppose that his money has no cost but in fact the opportunity cost is whatever would have been the return on that money in his next most preferred investment. This might be a bank deposit account or the stock exchange. If he could have invested at 6 per cent then the opportunity cost of investment in his own holding is the 6 per cent foregone.

The opportunity cost principle is a very useful guide in the purchasing of inputs, even though by name it is seldom uppermost in the minds of the decision makers. It can also be applied when choosing between alternative crops.

Time

It is most important to define the length of time over which costs are incurred although time itself produces no effects. *Short term* and *long term* planning periods are usually referred to. The short term is a period within which changes in output can be made without altering the size of the production unit or plant. A vegetable grower might increase yields by using more fertiliser and improving his crop protection measures within a single season. He will do this at extra cost but the area of vegetables grown remains unchanged. Over several seasons, by contrast, the vegetable grower could expand or contract the acreage grown or the number of crops. The long term therefore refers to a period over which it is possible to change the production unit or plant.

Fixed and variable costs

In addition to identifying two time periods we talk of two types of cost *fixed* or common and *variable* or direct. A cost incurred regardless of the volume of production, even if the holding is left fallow, is a fixed cost. It has no bearing on decisions to expand or contract the area of crop grown and only becomes fixed after it has been incurred. A grower will have to pay regularly employed men, rates, rents, obsolescence depreciation and telephone charges, all typical fixed costs, regardless of whether he grows crops or not. Variable costs are directly associated with the level of production; they are the costs of inputs used in increasing or decreasing

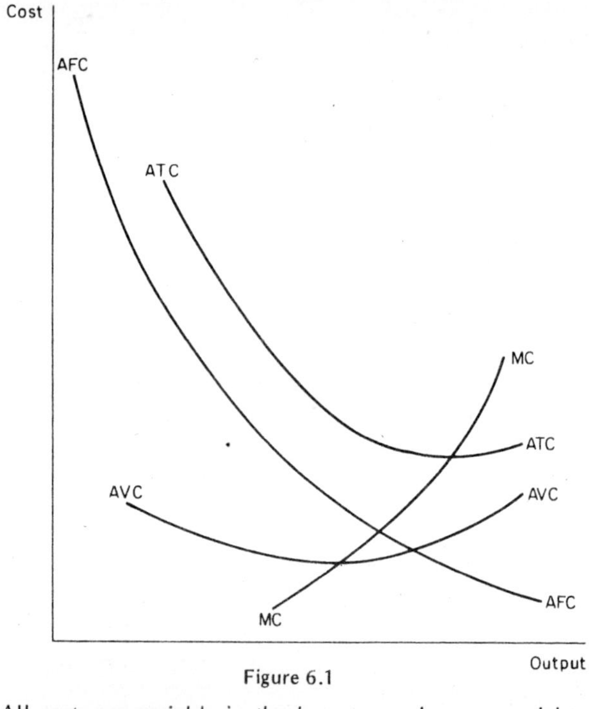

Figure 6.1

output. All costs are variable in the long term. A grower might use more seed, fertiliser, spray materials or casual labour, all typical variable costs directly related to increasing output. As variable costs are pertinent to decision making they are sometimes known as the *relevant* costs. The costs that are relevant for a particular decision depend upon the length of the planning period and the nature of the decision. A fruit grower who has committed himself to long term cropping might consider that all costs in the production of a single crop are fixed. He may on the other hand, only regard picking costs as a variable depending upon the yield of fruit. As a further alternative he might even treat the crops in specific years as annual crops and all associated expenditure as being variable.

Total, average and marginal costs

The sum of the fixed and variable costs is the total cost. It is the difference between total revenue and total costs that determines the net revenue or profits and whether growers will stay in production in the long term. As production is increased extra cost is incurred so that the total cost increases with output. The total cost curve is determined by the production function. In chapter 4 we discussed the production function in terms of physical components but these are easily converted to financial units by multiplying the physical values by the price paid for their use. The cost of inputs is always related to an output.

An *average cost* is derived by dividing the total cost by the number of units of output. *Marginal cost* refers to the increase in cost of producing an extra unit of output, or put another way, the cost that would not have been incurred if the extra unit had not been produced.

Table 6.1

Cost structure for runner beans (Phaseolus multiflorus) on a holding in England, 1967

Yield Tons	*Variable costs (VC)* £	*Fixed costs (FC)* £	*Total costs (TC)* £	*Average variable costs (AVC)* £	*Average fixed costs (AFC)* £	*Average total costs (ATC)* £	*Marginal cost (MC)* £
4	213	20	233	53	5	58.3	
							4
5	217	20	237	43	4	47.4	
							6
6	223	20	243	37	3.3	40.5	
							19
7	242	20	262	35	2.9	37.4	
							30
8	272	20	292	34	2.5	36.5	
							46
9	328	20	348	36	2.2	38.7	

The typical relationship between the total, average and marginal cost curves for varying output is illustrated in figure 6.1. The average fixed cost (AFC) or fixed cost per unit of output is downward sloping from left to right and falls continuously because fixed costs are independent of output. Variable costs are dependent upon output and the average variable cost (AVC) or variable cost per unit of output has a characteristic 'U' shape because diminishing marginal returns occur. To determine the exact shape empirical investigation is necessary for particular situations. The average total cost (ATC) is the vertical sum of the AVC and AFC curves. The marginal cost (MC) curve cuts the AVC and ATC curves at their minimum points.

In practice the relationships are likely to be more complex; the ATC curve, for example, may be a very narrow 'U' or more like a flat-bottomed boat in shape. The data in table 6.1 refer to runner bean production on a holding in England and are presented as an example. It is suggested that the student not only checks the calculation of average and marginal costs but also plots the data for himself to test the validity of the relationships in figure 6.1. Remember to observe the same rules as in chapter 4 when plotting the marginal cost.

Cost and output

We have stressed the importance of relating costs to output several times yet growers are commonly heard discussing the necessity of reducing costs to increase profitability. This may be commendable if output and price remain unchanged but they may well find that by divorcing the two, irrational decisions result from under-utilisation of resources. Table 6.2 relates the cost of fertiliser, yield and revenue associated with blackcurrant growing in England. It is apparent that there is a response to spending on fertiliser and that some growers may not be using enough. It may be worthwhile to *increase* cost to expand output and net revenue.

Table 6.2

Comparison of fertiliser cost and yields of blackcurrants (Ribes nigrum) in England (1958-59)

Total cost of fertilisers per acre for 2 years	£ 0-25	£ 25-50	£ 50-75	£ Over 75
Average yield per acre per year	Tons 1.17	Tons 1.61	Tons 2.37	Tons 3.13
Average revenue per acre per year at £135 per ton	£ 158	£ 217	£ 320	£ 423

Rendell, J. (1961). 'Blackcurrant Survey. Three Year Report, 1957-59', University of Bristol.

Scale economies

In the short term each firm has an average total cost curve representing its particular costs for a given output. An associated marginal cost curve cuts the curve at its minimum, this being the level of minimum cost production. In the longer term however, the producer can vary more costs and the size of his output. In figure 6.2, which is a very hypothetical situation, three firms are considered and the short run average cost curve (SRAC) and marginal cost curve (SRMC) have been drawn for each. The long run average cost curve (LRAC) is a tangent to these curves and has also been plotted. It is a shallower and longer 'U' than the short run curves because there are more variable costs. At its minimum point it is equal to the long run marginal cost curve (LRMC) and this is also the long run minimum cost level.

Analysing the diagram further we can say that costs fall as output is increased up to the point Q_1; after this cost increases again. Over the range OQ_1 we say that there are increasing returns to scale and beyond Q_1, decreasing returns to scale. Overall we may talk of *economies* or diseconomies *of scale.* They are particularly important to the manager seeking to reduce average unit costs with higher levels of output.

Statistical evidence from real industrial situations suggests that in most cases the curves are only downward sloping. This does not mean that the hypothetical situation depicted in figure 6.2 is impossible. In fact it is in industries like horticulture, dominated by small firms, that the upward movement of the long run average cost curve is most likely to occur. It may even be a reason why small units predominate. The general downward slope of the curves is probably due to firms becoming more efficient as they progress with a particular process or culture of a crop.

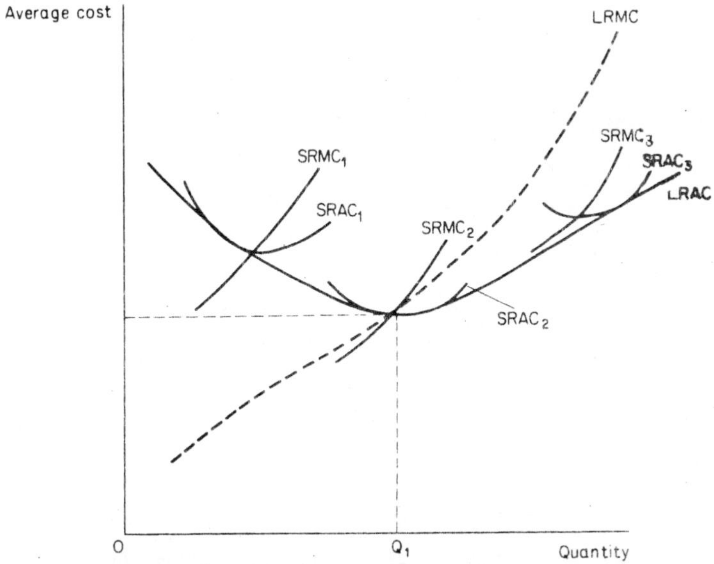

Figure 6.2

Other sources of economies and diseconomies of scale are less easy to identify. One source is best illustrated by the analogy of increasing the size of a storage tank. By using a greater area of material to make the walls of a cylindrical tank the volume can be enlarged. The capacity will increase at a faster rate than the surface area according to the geometric relation between them. If the volume is continually increased a point will be reached when extra structural members and supports will be required so that costs begin to rise. A second source of scale economy might be related to the range of sizes or capacities of machines available, many being produced in one size only. A grower may purchase a harvesting machine that requires an optimum of 20 hectares of crop to work efficiently. Up to this optimum the level of performance will increase but later will decline as the machine is unable to cope. Men and machines only exist as whole units so that in small firms employing few men an ideal distribution of labour between tasks is not always possible. A third source is in long production runs or mass production when economies can be achieved by automation, substituting capital for labour. If a sufficiently large area of crop is grown mechanical harvesting may be worthwhile and probably in addition, storage and transport facilities. Lastly managers themselves differ in their ability and for each there is an optimal size of firm that he can manage efficiently.

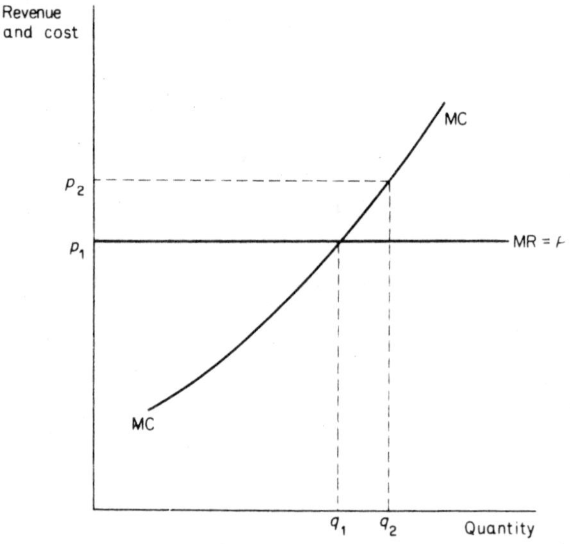

Figure 6.3

Measuring economies of scale in horticulture is very difficult. We could record and compare the performances of existing organisations but this raises the problem of obtaining a satisfactory representative sample of similar business situations. The resultant data still have all the inherent dangers that arise when historical data are used to predict the future. Another approach might be to construct hypothetical model situations but for this all too little information exists in horticulture. Some analysis of scale economies should be undertaken by the discerning manager. It is very relevant, for example, to the purchasing of buildings, glasshouses, packhouses, machinery, vehicles and the planting of orchards, or other crops.

The equilibrium of the firm and profit maximisation

We have seen in chapter 2 that as price rises and falls it is logical to expect change in supply. It is further reasonable to suppose that this will be reflected in output and the marginal cost involved in its production. The amount of revenue will also vary according to market prices. In considering an increase in marginal cost the grower will consider the likely increase in revenue, the *marginal revenue* (MR). The astute student will probably have already noticed that in the case of the horticultural free market, the marginal revenue is the same as the price, MR = p. If one box of apples is sent to market and realises £1 and later two boxes are sent and realise £1.20 each, the marginal revenue has increased from £1 to £1.20. Figure 6.3 represents the interaction of marginal cost and marginal revenue. At the point of intersection MC = MR. If the quantity of output is increased beyond this, then MC exceeds MR and below this more revenue could always be obtained by increasing marginal cost. *At this point the firm is maximising its profit.*

If the price at the market increases from p_1 to p_2 then it will be worthwhile for the firm to expand production to q_2. Alternatively, a firm having a marginal cost curve as illustrated can identify the appropriate size of output to aim for at q_1 by equating marginal cost and marginal revenue. Just as the market equilibrium for the whole industry is determined by supply and demand so the firm's equilibrium has marginal cost and price as components.

6.2 DETERMINING THE QUANTITY OF RESOURCES TO USE

In organising his production the manager is able to use capital, land and labour in varying proportions and achieve very differing results. He is most interested to know the best profitable combination and the rules for achieving this. To evaluate these we must return to the production function concept introduced in chapter 4. There we considered a single input X_1 in relation to output Y_1. This is often called a resource-product relationship by economists. In this chapter we are concerned with choosing between several resources; these are called resource-resource relationships. Later in this chapter we shall consider the choice between products or crops called product-product relationships.

The isoproduct curve

In practice the grower may have to choose between many alternating combinations of resources but only two variables can be considered and interpreted in graphical analysis. We shall see later that this is quite adequate to establish the guiding principles in resource allocation no matter how many are involved. Our analysis is concerned with the level of output Y produced from different combinations of two resources X_1 and X_2. The curve YY in figure 6.4 represents the outcome of possible combinations of X_1 and X_2. At every point on the curve the output is identical and for this reason it is called the *isoproduct* or *isoquant* (iso means equal). X_1 and X_2 might be labour and capital respectively or in more technical terms be seed and fertiliser or two fertilisers with respect to a single crop and area of land.

A greater level of output than YY is represented by other isoproduct curves farther from the origin. In fact there can be any number of isoproduct curves, YY, Y_1Y_1, etc. in figure 6.4. The curves cannot cross and are convex to the origin.

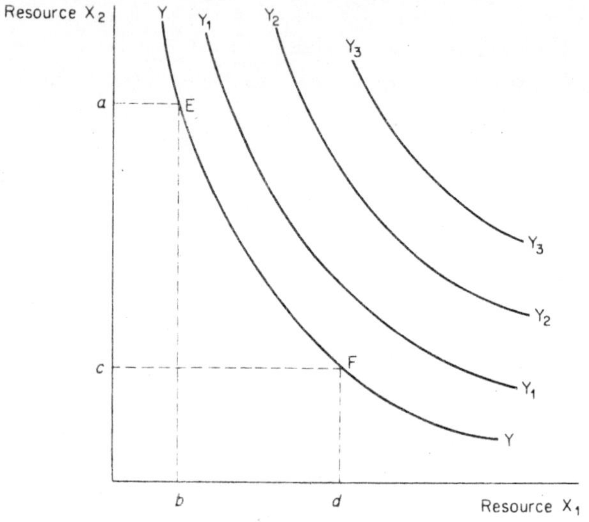

Figure 6.4

The isoproduct curve for a given output also represents the interchangeability of X_1 for X_2, *a* units of X_2 together with *b* units of X_1 would yield the same product as *c* units of X_2 and *d* of X_1. Thus, as we move from E to F there is an interchange of resources which is called substitution; we encountered this concept in chapter 2 with respect to indifference curve analysis. The rate of substitution indicates the amount by which X_1 must be changed to offset a change in X_2 to maintain the same level of output. As it is concerned with using extra or fewer units of resource we talk of the *marginal rate of substitution.* This is the rate of change of X_2 divided by the rate of change of X_1 or $\Delta X_2/\Delta X_1$; it is illustrated by hypothetical data in table 6.3. The student should again check the arithmetic and plot the isoproduct curve. By now even the non mathematically minded student will have noticed that $\Delta X_2/\Delta X_1$ is the slope of the curve.

If the cost of the relative resources is known we can calculate the cost of combinations of X_1 and X_2 and this has been done in the fourth column

Table 6.3

The least cost combination of inputs

Resources X_1	X_2 (Physical Units)	$\Delta X_2/\Delta X_1$	Cost† £
1	14	–	29
2	12	2	26
3	10	2	23
5	7	1.5	19
7	5	1	17
10	3	0.66	16
13	2	0.33	17

† PX_1 = £1, PX_2 = £2

of table 6.3. Cost falls to an apparent minimum in the table at 10 units of X_1 and 3 of X_2 and then rises again. We would expect the absolute minimum to lie very close to this on the isoproduct curve. It is impossible to be precise because of the units used. If however we plot the isoproduct curve as in figure 6.5, we can confirm the point of minimum cost. *Isocost* lines have been superimposed on the diagram and represent the relationship between the cost of X_1 and X_2; that is £1 and £2 respectively or a ratio of 1:2. One of these lines will just touch the isoproduct curve; at

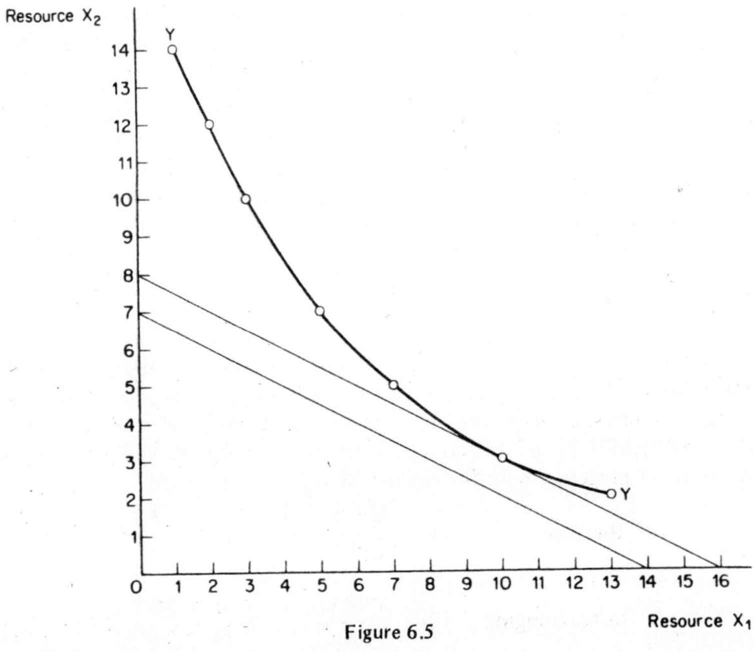

Figure 6.5

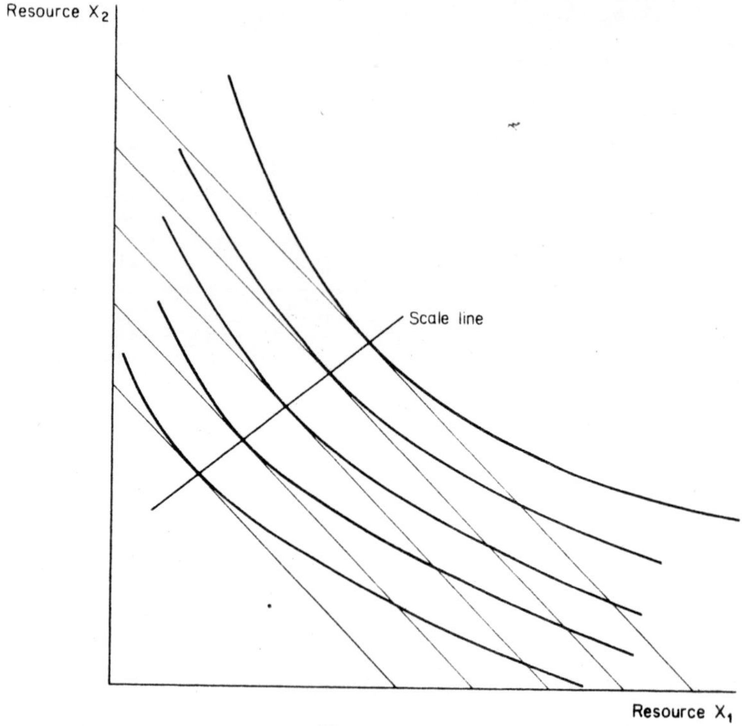

Figure 6.6

that point their slopes are equal and this is the point of minimum cost. If there is a whole family of isoproduct curves each with associated isocost lines, the points of tangency represent the levels of minimum cost or rational production. A line joining these points is called the *scale line* and describes the expansion path of a producer increasing his output as shown in figure 6.6.

Marginal substitution

We have established that at the point of tangency of the isocost and isoproduct curves, $\Delta X_2/\Delta X_1 = PX_1/PX_2$. At that point the marginal rate of substitution of X_1 for X_2 is also equal to the marginal product and as the product is physical it is known as the *marginal physical product* (MPP). The $MPPX_1/MPPX_2$ is also equal to the slope of the isoproduct curve at the point of tangency with the isocost curve.

therefore (1) $$\frac{MPPX_1}{MPPX_2} = \frac{PX_1}{PX_2}$$

or rearranging (2) $$\frac{MPPX_1}{PX_1} = \frac{MPPX_2}{PX_2}$$

In reality, a manager contemplating a change in resource use will ask himself if this will yield more than it costs. In the same manner we can substitute in equation (2) above, the price that we must pay for the use of X_1 and X_2 and assess the relative resources in use now and those that it is proposed to use. If the cost of the resource we are going to replace is greater than the cost of the resource we are proposing to use the costs will be reduced if we go ahead with the substitution. In our equation (2) if $MPPX_1/PX_1$ is less than $MPPX_2/PX_2$ it will be worthwhile using more X_1 until the equation just balances. At this point no further substitution will be worthwhile. Suppose for example that in equation (2) X_1 is a machine and X_2 manpower then at the point when the equation just balances the same amount spent on each yields the same product. The reader can verify this principle for himself by substituting numbers into the ratios and carrying out trial substitutions. Suppose, for example, that numbers are introduced as follows, both situations producing the same total output.

$$\frac{MPPX_1}{PX_1} = \frac{10}{6} \quad \text{and} \quad \frac{MPPX_2}{PX_2} = \frac{12}{6}$$

The equation does not balance and efficiency in production is not optimal. Resource substitution will be necessary to achieve an optimal use of X_1 and X_2.

The manager may also be concerned to know how he can best allocate his spending between resources. In answer we can say that he should distribute his outlay between X_1 and X_2 such that the MPP of a single unit of X_1 equals the MPP of the same amount of X_2. These simple principles apply however many resources are considered.

The demand for resources

Substitution between resources is a continuous process and the rate for a particular product will depend upon the relative cost or price paid for the resources by the manager and on their productivity. Not surprisingly therefore, the quantity of resources used by our agricultural firm varies with time. The demand by producers for a resource can be measured with respect to price in the same manner as we discussed for consumer demand in chapter 2. The mathematics involved in the analysis can however be very complex and are beyond the scope of this book. This type of analysis has not been made for horticulture due to lack of useful data. Table 6.4 is presented as an example of some estimated elasticities of aggregate demand for resources with respect to UK agriculture in 1965-66 and were made by a worker at the University of Manchester. They are expressed in terms of product prices in two time periods. The supply of land and farmer's labour is assumed to be fixed. Hired labour and the feed, seed and sundries category, both show a response to price but neither changes with time. If for example, the price of hired labour increases by 1 per cent then the demand for it will increase by 0.86 per cent. With respect to fertilisers, machinery and buildings however, farmers show a greater response in the longer term than the shorter.

Table 6.4

Elasticity of demand with respect to product price. UK agriculture, 1965-66

	Short run	*Long run*
Hired labour	0.86	0.86
Fertiliser	0.70	1.00
Feed, seed and sundries	0.30	0.30
Machinery	0.24	0.35
Land	0.00	0.00
Buildings	0.25	0.62
Labour of farmer	0.00	0.00

Rayner, A. J. (1970). 'Demand for inputs and the aggregate supply function', *Journal of Agricultural Economics*, 21, 233.

Irrational production

The irrational use of inputs was discussed in chapter 4 with respect to the production function but it can also be demonstrated in the isoproduct curve. In figure 6.7 the rational producer would only use the range TS of the isoproduct curve YY. At all other points on the curve the resources are not substitutes but actually complement each other, so that more resources are needed to produce the same product than in the range TS. A producer at f for example, requires *e* units of X_2 and *g* units of X_1 to produce the same output YY as can be produced over the range Oc and Ob, respectively. Production outside the range TS therefore, represents irrational production.

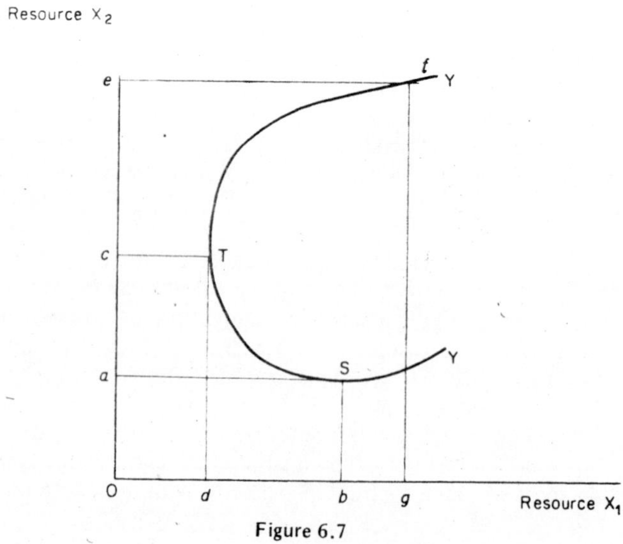

Figure 6.7

6.3 DETERMINING WHICH PRODUCT OR CROP TO GROW

Product-product relationships are concerned with the choosing of the best combination of crops or enterprises from an available stock of resources. The principles involved in optimal choice are very similar to those discussed in the last section.

Competitive products

For simplicity, we again use a two factor model Y_1 and Y_2 figure 6.8(a) illustrating a typical choice between two crops. Only one crop can be grown on a given area of land at a time and so they have a competitive relationship for this resource, an increase in production of one requiring a decrease of the other. This is the normal situation encountered in horticulture. The curve GH represents the maximum amount of output that can be produced for particular levels of Y_1 and Y_2. The slope of the curve or its rate of change describes the now familiar substitution relation between Y_1 and Y_2. A grower's production possibilities could be, for example, *Oa* units of Y_2 and *Ob* units of Y_1. Alternatively he might grow *Oc* units of Y_2 and *Od* units of Y_1. The units are usually land area.

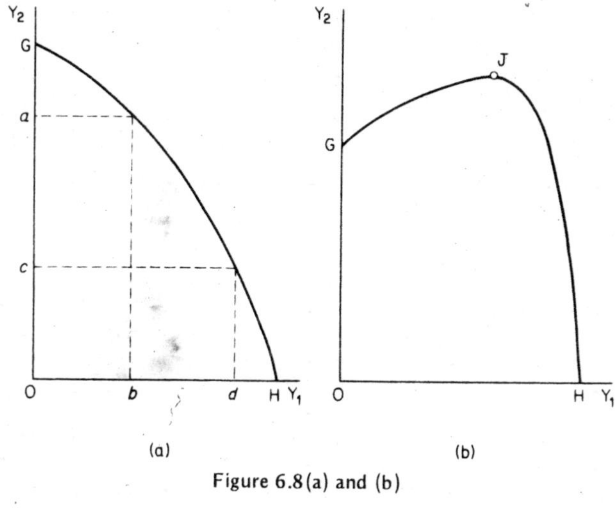

Figure 6.8(a) and (b)

Complementary products

Over the curve, GJ in figure 6.8(b) a grower could increase the area of both Y_1 and Y_2 but at J the products become competitive again. There is a *complementary* relationship between enterprises. Increasing production of Y_1 has added to the production of Y_2 even though resources have been transferred from one to the other. The relationships never hold over the whole curve, always becoming competitive at some point.

Perhaps the best example of this type of relationship is the use of rotations particularly where they are used to ameliorate the soil. The use of short term grass leys to improve soil structure is one example; another is the nitrogen fixing properties of the nodulating legumes, the benefits in both of these cases being passed on to succeeding crops. This illustrates another feature of complementary products: they can only occur over several production periods or crops and not within a single cropping period.

Supplementary products

If we could increase the production of Y_1 without altering the production of Y_2 then the products are said to be *supplementary*. This is illustrated in figure 6.9(a) where GK is the range of suppletive potential, the curves becoming competitive at K.

The use of grassed orchards as a source of grazing might be an example. Part time farming and 'catch cropping' and 'inter cropping' in horticulture may be further examples according to how they are organised. For all these examples the point where they become competitive with the main enterprise is easy to see. Thus inter crops compete when there is interaction between the crops due to plant competition for water, nutrient and light.

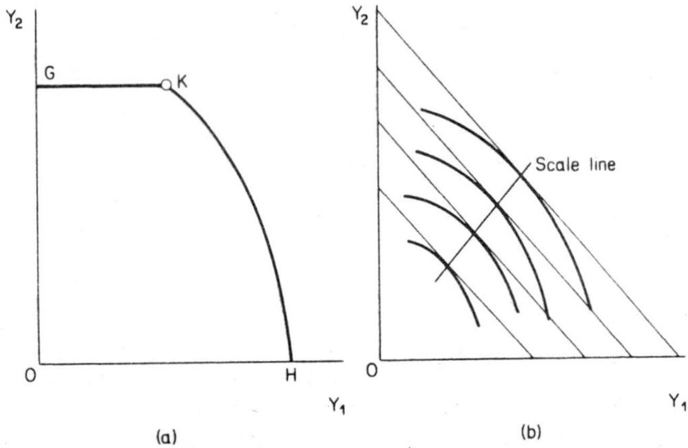

Figure 6.9(a) and (b)

The optimal combination of crops

The principles involved in selecting the most profitable combination of crops are similar to those for resource use. The characteristics of the production possibility curves are also similar. Families of curves may exist, representing higher levels of total product the farther they are from the origin. *Isorevenue* curves can be superimposed and the points of tangency of these with the production possibility curves is equal to their points of highest revenue. Joining up those points gives an expansion or scale line: figure 6.9(b). Note that we are no longer concerned with physical product but with a product having value. In the same way as yield multiplied by price gives the total gross revenue so marginal physical product of Y_1 multiplied by the price obtained for Y_1 equals the *marginal value product,* $MVPY_1$. The student will recall the equation evolved in section 6.2 for optimal resource use and if we multiply through this by the price of Y_1 and Y_2 it becomes

$$(3) \quad \frac{MVPxY_1}{Px} = \frac{MVPxY_2}{Px}$$

Given a stock of resources (x) the grower may wish to know which crop he should grow. It will be worthwhile for him to substitute Y_2 for Y_1 as long as the $MVPY_2$ exceeds the price that we pay for the resources used in production. We are equating extra revenue, the MVP and marginal cost of the resources. This as we have already established is a profit maximising situation. If Y_1 were cauliflower and Y_2 were brussels sprouts then at the point when the equation exactly balanced there would be no incentive to substitute one for the other.

The principle of resource and enterprise allocation that we have considered is often called the *principle of equimarginal returns.* Formally defined it says that the value productivity of the last unit of any one resource employed in producing any product must be equal for all products. We shall see later how this principle is the basis of modern planning techniques.

Unfortunately there are some practical limitations to the universal use and interpretation of the principle of equimarginal returns. Firstly, in horticulture there are too few accumulated data on the relationships between resources and enterprises. Secondly, capital is the most commonly limiting resource impeding the optimal combination of inputs. Thirdly, most growers err on the 'safe-side' in resource use and do not commit resources fully as a 'hedge' against uncertainty. Fourthly, inputs are often indivisible so that a perfect solution is impossible, the most obvious example being labour which can only be considered in terms of whole individuals. Lastly, a further difficulty of labour allocation is that its productivity may change in time. For example, a man may be more productive in summer than winter or be variable according to his age.

6.4 PLANNING TECHNIQUES

The overall profitability of the business depends upon the combination of crops that are grown. The quantity of resources available for any grower will be limiting at some level. Thus he may have only 20 hectares of land and that not of equal fertility, together with a limited amount of capital and only two men as a labour force. His limited resources restrict the output.

Consider a hypothetical example of two crops Y_1 and Y_2 in figure

6.10 competitive for a grower's resources. Let us further assume that Y_1 requires space in the propagation house but the firm has limited facilities A. Y_2 by contrast has no requirement of propagation space. The line AB therefore, represents the production possibility in terms of propagation facilities. Suppose each crop requires land and the line CD represents the relative land constraint. Labour is plentifully available except in June when it becomes limiting for Y_2 at E but Y_1 has no requirement of June labour and therefore the line EF represents the June labour constraint. In practice many more constraints may exist but even in this very simplified model a production possibility curve has been delineated by the boundary EGHA.

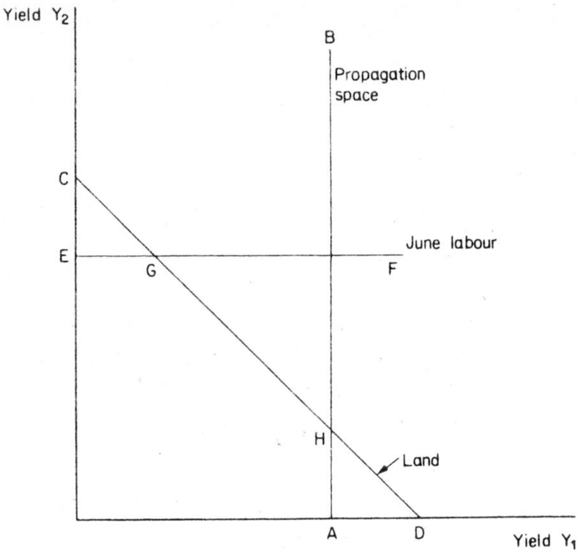

Figure 6.10

In all planning we are concerned with input-output relationships. It is useful therefore to establish indicators of these relationships in terms of the potential profitability of specific crops. One such indicator is the so called *gross margin*, which is the excess of gross revenue over the variable costs of production. Each crop has associated variable costs and a gross margin. The sum of the gross margins of all the crops grown gives the gross margin generated for the whole holding. If the total fixed cost is deducted from this the profits of the holding remain. Furthermore, we can say that the higher the gross margin generated by a particular crop the larger will be its contribution towards paying the fixed costs of the holding. In planning for the future the higher the gross margin foreseen the greater will be the potential profit.

Gross margins are simple to understand and are commonly used. Since they are calculated from the yield, price and variable cost they contain all the pertinent variables for effective decision making and planning. There are many planning techniques available to the manager but here we discuss only the three main methods used in horticulture.

Budgeting

The process by which we estimate the likely profitability from the use of resources in growing crops in different combinations is called budgeting. Essentially this tool of management is a statement on paper of the financial and physical implications of proposed policies with respect to the horticultural business. In practice, it may be done in many ways from the back of a cigarette packet to highly sophisticated forms provided by extension officers. The author has even seen it done on the paintwork of a glasshouse door! The important thing is that it is done, for it makes the grower think about his future actions and plan them.

The manager sets down the likely cost of resources to be used in relation to probable returns. The main problem arising here, apart from uncertainty of the outcome, is the feasibility of the figures used for a particular business. Ideally they should be derived from the holding but if the crop has never been grown before, or records not kept, then data from surveys, case studies and experiments will have to be used. These sources have a hidden advantage in that they require the grower to study alternatives and to read the available literature. Prevailing or projected prices are used which is a weakness of all budgeting schemes because they do not necessarily hold in the future. Indeed by increasing the production of a crop the grower himself may influence the market price. For horticultural products sold in nearly pure markets this is not as serious as it may at first seem. There are two types of budget used in practice, complete and partial. The former refers to a complete plan for the holding, the latter to a small part of the business, such as the potential for changing one crop for another, or building a new glasshouse. For details of the layout and mechanics of budgeting the reader is referred to elementary texts on management. In summary however the procedure for a whole holding might be as follows. Crops generating the highest gross margin are utilised taking into account the resource constraints. Attention will be given to labour use throughout the year to ensure that it does not exceed the amount available at any time. The budget is drawn up to satisfy the goal of the grower which might be maximum profit. After the plan has been put into operation it may be reappraised from time to time and substitution of resources or enterprises made if appropriate.

In selecting a budget for maximum profitability the grower will probably observe more economic principles than he knows by name. He will, for example, have decided how much fertiliser to apply to a particular crop and probably reconciled whether the marginal cost justified the marginal revenue from fertiliser use. Resource use within the limits of those available will have been taken into account in deciding which or what quantity of crops to grow. If he chooses the most profitable solution substitution will have been considered and rejected. Unfortunately the input-output relationships change with time and reassessment is necessary. Thus budgeting is used at the outset of business activities and for reappraising at a later date.

A weakness of budgeting is that it is limited by the capacity of the human mind. It is quite satisfactory where the choice of alternative enterprises is very limited but when they are multitudinous it may only reinforce guesses or ideas on the part of the grower and/or adviser. The subjective nature of the data used in the compilation and inaccuracies that creep in through calculation errors are further disadvantages. A technique which goes some way to achieving a more logical solution is called programme planning.

Programme planning

Although somewhat involved, programme planning is a manual technique that can be carried out by grower or adviser with pencil and paper. Limiting resources are identified and the enterprises ranked in order of magnitude of the gross margins that they generate. The crop giving the highest gross margin is allocated land within the holding until a resource becomes limiting. Consideration is then given to obtaining more of the resource or using another enterprise which is more productive in the use of the limiting resource. Substitution of enterprises is made to utilise the available resources effectively even though a progressively lower gross margin may be generated. The final solution will eventually be reached when no further substitution can be made.

The final solution may not allocate all the resources completely, due to indivisibility of the units of resource and enterprise. Any plan involving more than say ten enterprises and resource constraints becomes tedious to handle and errors of calculation become easier to make and more difficult to identify. It is however a very much less costly technique than the third planning technique that we shall discuss – linear programming.

Linear programming

The mathematical procedures involved in making a linear programme are beyond the scope of this book. In principle it uses the same data as programme planning; all available resources and constraints of production are identified together with possible enterprises and their associated gross margins. The highest gross margin is selected from amongst the production possibilities and substitution is made thereafter until the resources are used as completely as possible. As its name implies, linearity, a straight line relationship, is assumed between inputs and outputs. This is rarely true in practice but the assumption is made to simplify the calculations and may be a good approximation. To minimise this limitation it is possible in the plan to reduce the amount of crop that can be grown for a particular quantity range of resources to say 10 – 20 hectares instead of 0 – 100 hectares.

Linear programming further assumes that inputs and outputs are perfectly divisable and might produce a solution suggesting for example, the growing of 0.44371 hectares of tomatoes and 0.55629 hectares of cucumbers. Fitting this exactly into a hectare of glass is clearly impractical and in reality the nearest whole unit is chosen. The solution will however allocate resources very much more completely than does programme planning.

The calculations are usually done with the assistance of an electronic computer which is really only a high speed calculating machine with the ability to store information. It is very quick and accurate, these qualities permitting the handling of many enterprises, resources and constraints which would be very time-consuming and liable to error if done manually. It can only produce a sound solution from accurate and correctly assembled data.

It is possible to vary the type of solution sought. Instead of maximum profitability we might for example programme the computer to produce a solution in terms of minimum loss. New planning techniques are continually being evolved; integer programming has been developed to cope with the problem of indivisibility of resources and enterprises; Monte Carlo programming produces several random solutions from combinations of inputs, from which a choice can be made according to personal requirements.

None of these techniques is likely to be used by growers themselves although model solutions for groups of homogenous holdings can be drawn up. Computer time may be expensive and the skills and abilities to operate them are still confined to researchers, specialist consultants and a few extension workers. Linear programming and programme planning are also invaluable teaching tools because both are based on sound economic principles.

In this chapter we have considered the basic decisions of what, and how much, to produce and we have reviewed the principles and techniques used in practice. One further decision remains for the grower, that of where to carry out his production and marketing. This has intra- and international implications and is the subject of the next chapter.

CHAPTER 7

LOCATION

Location is concerned with the geographical position of production units. Imagine that you are holding a near perfect sphere in your hand, say a ball-bearing. You can see the face nearest to you but if you throw it in the air and catch it again it is impossible to identify any point again because the whole surface looks alike. If we make a mark on the surface however, whatever we do with the sphere we can always define any other point in relation to the fixed mark. In horticultural locational studies we are also concerned with relative positions, those of horticultural holdings relative to markets at home and overseas and to the areas most favourable for production.

As the subject is about spatial distribution on the world's surface it is compounded of quantitative geography and agricultural economics.

7.1 INTRANATIONAL TRADE

A grower might be concerned as to where he should establish his holding. In practice this decision is not as commonly encountered as might be expected because often the business is inherited, evolves out of a parent firm, or is purchased without much thought being given to locational principles. More commonly the grower is forced to consider in his every day decision making what advantages or disadvantages his holding has with respect to production potential and market outlets. As outsiders we might view a horticultural production area and ponder as to why it was established there and why it specialises in that type of production. There is however no single locational problem for everything is relative and depends upon the prevailing social and economic environments.

Locational theory

The fundamental locational theory was developed in the early nineteenth century and is very simple in concept. It says that production will or should occur where the relative costs are lowest. This principle was developed before the advent of transport and market systems as we know them today and it does not consider the function of price in determining what is to be produced. Taking these factors into account economic theorists have evolved two very important locational principles, those of *absolute* and *comparative advantage.*

The climate of an area may make fruit-growing well worthwhile whereas another area may be best suited for vegetable production. Both crops could be grown in either area but in the area where they grow best each has an absolute advantage over the other. Growers of other crops may also gain advantages by producing in either area. If however only vegetable and fruit growers compete for land in both areas then they will have a comparative advantage as well in their respective areas. This is because although both compete for land, labour and capital they achieve different results from the use of them. The natural advantages in each area translate themselves into higher yields and prices for superior quality and lower costs of production. We return to this important principle later in this chapter when we discuss principles of international trade.

Agricultural locational theory has been developed by several workers but is still firmly based on the least cost principle even though today we must consider prices and transportation. An early nineteenth century German pioneer in locational studies, J.H. von Thünen, in a book called *The Isolated State*, showed that a zone devoted to perishable crops of high value would of necessity develop around a market and centre of consumption. He postulated the existence of zonal, concentric rings radiating outwards, each having a decreasingly intensive agriculture. In the days when transport was primitive this principle probably made more sense than it does now but examples can still be found. It is certainly common to find horticultural production located near urban centres. In some developing countries lack of transport is still a major locational limitation overriding climatic advantages for production elsewhere. In such places expansion of production may cease when local requirements are satisfied.

Locational advantages

There are many factors that may endow a production area with economic advantages but only a few usually predominate. Natural advantage due to a favourable climate, seasons or soils are obvious examples for the horticulturist. Most horticultural production is on highly productive land but many growers have suffered the results of poor location on inferior soils. Closely linked to climate may be the incidence of particular pests or pathogens. For example, one area may be more favourable than another for apple production because of a low incidence of apple scab *(Venturia inequalis)*.

Another natural advantage may result from a holding being located close to a market. To the grower, processor, wholesaler or retailer considering transportation today, the ratios of bulk to value and to distance are important. Close proximity to market and urban centre may confer a further advantage in providing a source of casual labour. Conversely, such proximity will probably lead to competition for workers and higher wages and may also pose problems of trespass, theft and vandalism. Land values will also be increased particularly where there is potential for development. It may become worthwhile to sell the holding and set up business again on another site using the capital created by the sale. This has happened in the UK particularly in the Middlesex and Lea Valley areas where the expansion of London has encroached on horticultural production. Many growers have been able to relocate themselves on improved production sites near the south coast. This type of migration may even be furthered by Government action in some countries when grants are provided to develop particular areas.

Once a centre of production becomes established its very presence has an accelerating effect on the growth of the area. Other potential growers seeing the benefits associated with the area are encouraged to move. An extension service, private or public, may evolve and firms selling horticultural sundries become prominent as a result of the potential business. In addition, roads and railways may develop to accommodate the needs of the industry so that the area becomes even more favourable for production. The holdings themselves may disintegrate in the economic sense; employees or members of the grower's family may go off and start their own holdings. Growers may specialise their production or even expand by starting another separate business. An example of this is the emergence of specialist propagators of chrysanthemums, carnations and poinsettias from cut-flower nurseries.

The more outstanding an advantage is to an industry in a particular area the greater will be the predominance of that industry. There will also be a narrow range of type of production in that area. A major disadvantage that can arise from a high degree of specialisation and localisation is that the whole area may become threatened if only a single but crucial locational factor changes. The closure of a local branch railway line vital for transport may have serious effects on a localised industry. Equally the building of a new motorway in an area hitherto remote or congested can raise the grower's marketing horizons.

It must be repeated that there is no single locational problem or solution. For this reason two case studies are presented at the end of the chapter illustrating some of the important features. In the meantime the student could usefully consider any horticultural holding known to him and evaluate the locational features which seem to make for advantage or disadvantage.

7.2 INTERNATIONAL TRADE

The origin of trade

Trade between countries occurs because resources are not equally distributed throughout the world. That geological features and climate vary throughout the world is well known but skills in production and organisation may also become locally developed. The emergence of exchange economies from self-sufficient peasant communities in the transitional phase of growth, often leads to trade between countries instead of being solely intranational. Indeed it only becomes international trade because of the artificial political boundaries delineated by man in his creation of uniquely bounded countries. Historically these boundaries were crossed by early travellers and explorers who recognised the potential in other countries for the use of new resources. Furthermore, their initiative in developing this trade was rewarded by the very scarcity of the goods. Transport was important even to this early trade and it often developed along natural waterways or sheltered valleys, routes which persisted for many years.

Technological improvement by breeding and cultural techniques was influential later, not only in enabling surpluses of agricultural produce to be generated for export, but also in providing the means of their transportation. Horse, rail and road transport, followed by refrigeration and packaging, have all contributed in turn to the development of trade in perishable agricultural products. In more recent years national agencies have emerged to export on behalf of growers and countries have cooperated to form free trading areas. The most recent influence on international trade in horticulture is the increasing use of air transport. There are now very few undiscovered or even inaccessible places left in the world.

The theory of comparative advantage

International trade differs from intranational in several fundamental ways. It is concerned with the passage of goods across man-made political frontiers and the barriers imposed by respective governments. The economic organisation of the countries may not be the same. Capital may well be tied to particular currency areas and language differences may be a further barrier. Immigration restrictions may affect movement of labour, the productivity of which may differ from country to country. For these reasons a separate theory is usually stated to describe the principles by which international trade is organised. This is done even though there are obvious similarities in the principles of trade or exchange of goods whether within a country or between countries.

Comparative and absolute advantages are again the guiding principles. Trade would logically occur when the cost, in money terms, is lower in one country than another. We would expect produce to move from the low cost to the high cost country. This is too simple and needless to say not the whole story. Money is paid for the use of the factors of production, land, labour and capital, and the *factor cost* of producing refers to the quantity of these factors used in producing the goods. It is reasonable to suppose that any country will try to use these resources where they are most productive. A country may well be able to produce for example, apples and citrus but can produce more apples for a given amount of resources than it can citrus. It will clearly do better to specialise in the production of apples. The perceptive student may have noticed the concept of opportunity cost that we discussed in chapter 6 in this example. It has determined the degree of specialisation in the respective countries because the opportunity costs are different and trade between them would be worthwhile.

Suppose that our two countries X and Y from the same amount of resources can produce as follows

	in country X	can produce	1000 units of apples
100 resource	in country X	can produce	600 units of citrus
units	in country Y	can produce	500 units of apples
	in country Y	can produce	2000 units of citrus

Clearly X has an absolute and comparative advantage over Y in the production of apples but Y over X in the production of citrus. The opportunity cost principle tells us that they will benefit from specialisation and trade, as both countries will save 100 units of resource by not having to produce the products for which they have a disadvantage. In reality the issue will be clouded by such considerations as transport costs and differential wage rates. Even if Y could produce only 500 units of citrus instead of 2000 it would still retain its comparative advantage over X for citrus; for 100 units of labour it could still produce equal amounts of apples and citrus whilst in X the ratio is 1000 : 600.

In summary the theory of comparative advantage (or cost) states that *two countries will gain by specialisation and trade provided that each has lower comparative costs in the commodities it exports.* This theory can be extended to any number of commodities and countries. It is important to remember that any advantage or disadvantage is changing continually; there is no fixed line between what a country imports or exports. In practice trade may not be a simple agreement between two countries (a bilateral agreement) either, but is commonly of the multilateral type where several countries are involved.

The terms of trade

The movement of goods internationally is called 'visible' trade and that of services 'invisible'. The latter include for example, tourism, banking and insurance. The sum of visibles and invisibles entering or leaving the country in a given time period represents the imports and exports of that country. The relative amounts of these are partly determined by what is called the terms of trade – the price advantage or disadvantage of trading. If there is a price increase in favour of one or another commodity then there will be a shift in the terms of trade. If export prices rise more or fall less than import prices, then the terms of trade are said to move 'in favour' of the exporting country. In the example used above a change in the price of apples relative to oranges, within either X or Y, would change the terms of trade between them.

The terms of trade are usually measured as an index using a base date and expressing subsequent values at the base date prices. For example if 1970 is a base year, we might say that exports in 1975 sold for 115 or conversely at 1970 prices they would have sold for 100. If the same thing is done for imports, then if imports were 120 we would say that the terms of trade have worsened by 5 per cent (120 – 115). Government offices usually publish import/export prices and the terms of trade; they change the base year from time to time to maintain meaningful comparisons. The terms of trade are frequently confused with the 'balance of payments' of trade. Although they are interconnected they are eventually different. The balance of trade is a comparison of total value whereas we have seen above that the terms of trade are concerned with the relationship between the prices of imports and exports.

World prices are largely decided by supply and demand. After making an allowance for the cost of transport, the world price of goods arriving at any port, at a moment in time, tends to be much the same everywhere. However on crossing the national frontier the price may be increased by importers' handling charges, profits and by import duties imposed by the importing country. These artificial constraints may prevent a rational price equalisation between the countries according to their monetary rate of exchange.

Restrictions on trade

Most countries today impose some form of restrictions on at least some of their imports. The best known of these is the tariff which is a tax levied on imports crossing the frontier; these are discussed more fully in chapter 8. Restrictions may also take the form of a quantity limitation (quota), be related to a specific time period (timetable), or there may even be a total ban on all imports (embargo). Plant health regulations that prevent the entry of certain types of produce may also interfere with trade.

There are several reasons to put forward for the protection of home industries by the restriction of imports, one of the commonest being the need to protect the standard of living of workers in high income countries by preventing the importation of goods from countries with lower production costs. In fact low cost imports in high income countries tend to raise the standard of living rather than the reverse. It is output per head, the share of the total product generated that determines the standard of living. Attempts to reduce the inequality of low and high income countries do not change the total to be shared between the population unless the low income countries are able to raise their productivity.

Another reason often given for protection is that it allows home producers competing with imports to expand their output. This may well be true but artificially maintaining prices makes the consuming population suffer. Furthermore the cost of living may rise generally throughout the country reducing the spending value of wages. For developing and sometimes other countries protection is often deemed necessary to nurture small or new industries before they are exposed to the full force of competition. The value of protecting these so called 'infant industries' is not proven. A last example is that of a country choosing to maintain a large rural population and deliberately protecting its home agriculture towards this end.

The problem of the developing countries

Protectionist policies may delay access to the markets of the developed countries by exports from developing countries. This is the central problem concerning the trade of the developing countries and is particularly crucial in their transitional phase of development. They have recently been increasing the volume of exports of food by 2.5 per cent per year but have received declining prices. In the same period demand in developed countries has increased by 4.4 per cent and it is the exports from other developed countries that have supplied the difference. The result is that the rich get richer and the poor get poorer, in relative terms. Worse still, some developing countries may even be substantial importers of food. India and Pakistan in poor crop years; Hong Kong and Singapore because of their limited size and Sri Lanka and Malaysia because their agriculture is highly organised for specialist export crops, are examples.

Primary products account for 90 per cent of the total combined foreign exchange of Africa, Asia and Latin America. These products are very susceptible in world markets particularly to technological advances stemming from the developed countries, making the traditional methods of production inefficient, or there may be a change in demand as consumers adopt new diets. Some may be affected by synthetic substitutes, for example, saccharine for sugar or plastics for rubber. As yet, international trade in horticultural products is little affected by this type of development.

Apart from the shorter term measure of giving aid in the form of food, cash or other resources there is only one logical long term way to really help these countries. This is to help them to increase their exports by the provision of better opportunities to sell in world markets so that funds can accumulate at home. This may also go some way to relieving poverty which is the real cause of hunger and malnutrition; the people will be able to buy food to eat.

Towards free trade

So-called free trade occurs in the absence of import and currency exchange restrictions, taxes and subsidies on production. In practice it is still quite rare. However some groups of countries have agreed to form what are called free trade associations in which they agree to protect themselves by imposing restrictions on imports at their common external frontier. Between themselves they progressively reduce trade barriers. Examples are the European Economic Community (EEC) and the Latin American (LAFTA) and Caribbean (CARIFTA) Free Trade Associations. These were formed in 1958, 1960 and 1968 respectively.

A few international institutions often associated with the United Nations have emerged during the past 25 years in recognition of the need to reduce trade barriers. The General Agreement on Tariffs and Trade (GATT) is one such organisation. Formed in 1950 its membership embraces approximately 80 per cent of world trade. Its main aim is to convert all forms of protection to tariffs and progressively reduce these by negotiation. In particular it aims to protect the interests of the lesser developed countries without damaging the trading interests of all parties. GATT has existed for over 20 years but progress has been painfully slow; however it has managed to weather its own problems and is still strongly supported by many influential governments. Attempts have been made from time to time to accelerate progress. The late President Kennedy, for example, in 1962 gained powers from the US Congress through a 'Trade Expansion Act' to make 50 per cent cuts in US tariffs over a period of five years. This depended upon other countries also making concessions and through GATT what came to be called the 'Kennedy Round' of negotiations got under way. They lasted four years and concessions were gained mainly on industrial products but not to the extent that had been hoped, horticulture hardly being affected. In the UK for example the tariff on gooseberries *(Ribes grossularia)* and raspberries *(Rubus idaeus)* has been reduced by 1 per cent and 3 per cent respectively. Neither crop is important in international trade and the reduction is hardly significant.

Another example of an organisation established to deal specifically with the problems of developing countries is the United Nations Conference on Trade and Development (UNCTAD). This was formed in 1964 and is a focal point for many new ideas but again progress is slow. The developed countries also have formed organisations to act as mirrors for all to see how others are progressing. One example is the Organisation for Economic Cooperation and Development (OECD) where membership includes all the countries of the European Free Trade Association (EFTA), the EEC as well as the USA, Canada and Japan. It was formed in 1961 and has given technical assistance not only to its own members but also to lesser developed countries.

7.3 CASE STUDIES OF LOCATION

Two case studies are presented to illustrate in practical terms some of the features of location discussed so far in this chapter. The first case is concerned with a small isolated horticultural area in the UK specialising in strawberry production. It is particularly concerned with intranational trade but in recent years has also been influenced by imports. International trade is the subject of the second case and takes the form of an investigation of factors that might relate to the exportation of apples from the USA to Brazil. As apples are produced in both countries there are also implications to producers. Whilst reading through the cases the student should attempt to identify the important locational factors.

CASE A. The Cheddar strawberry growing area

Locational features in development and production

Cheddar is an important tourist centre in the West of England very isolated from other horticulture and from the markets. It is a suitable area for horticulture because of its free draining neutral to acid soils which warm early. Good air drainage minimises the risk of frost and it is sheltered from northerly winds.

In the nineteenth century early potato production predominated. With the development of transport, particularly the railways in the latter part of the nineteenth century, other centres became stronger competitors for the potato markets. Coincident with this were the first recorded plantings of strawberries in the 1880s. Declining returns for potatoes and immediate financial success from strawberries led to the entry of more growers some new to horticulture and often working part time. So the strawberry acreage expanded and by the turn of the century most of the land that was available was under cultivation by which time transport to distant markets had developed. Special trains ran to the north of England and south of Scotland which became important outlets.

Today the industry is strongly localised and almost entirely contained in three parishes. Some 130 hectares of strawberries were cropped in 1970 on 210 hectares of holding. Over 60 per cent of the holdings are under 2 hectares in size; fragmentation is considerable and there is no land for further expansion. Family holdings predominate with most of them having a full-time working occupant with his family as the usual source of part-time labour.

Locational features in marketing

Transportation of strawberries to wholesale markets is still the normal method of marketing, only about a sixth of the growers selling a small proportion of their crop locally. They do this despite ever increasing transport costs and the fact that Cheddar is a much visited tourist area with considerable scope for extending retail trade.

The branch railway line serving Cheddar was closed a few years ago making the growers dependent upon road transport. The nearest railway station is 13 km away but there are two others, 22 and 35 km distant and to use these the growers must arrange road transport to the stations and further transport and handling is required at the destination station to get the produce into the markets. The railway has further decreased its service to growers by reducing the number of markets that are accessible from

Cheddar. Not surprisingly there has been a decline in number of packages handled by the railway.

Hired road transport is used to some extent by all growers to take produce to market. Few growers have or use their own transport but one grower does and carries the produce for some others. Wholesalers' vehicles collecting at the holdings account for about 14 per cent of produce transported, a service which grew up when the railway line was closed. More road transport problems have been encountered with the introduction of new laws restricting the use of certain sizes of vehicles and the number of hours the driver can be at the wheel. This has rendered some markets less accessible than before and further raised costs. Undoubtedly this will encourage further economies in transport use.

Foreseeable developments that will be important locational features are firstly the completion of a national motorway network which will shortly be extended to within a few miles of Cheddar. Secondly there is an anticipated expansion of Bristol and a new urban complex called 'Severnside' that will bring new consumers and locate them close to Cheddar. Thirdly the continued increase in tourist traffic in the area will raise still further the potential for local sales.

International competition

Cheddar is not an exceptionally early production area by UK standards but the advantages of advancing the strawberry season to seek higher prices are appreciated by Cheddar growers 90 per cent of whom use some form of protected cropping. By the late 1960s this produce began to be affected by overseas competition. The UK now receives imports of strawberries from late November until the start of the main home season in June. They come from New Zealand, Israel, Mexico, Cyprus, Kenya, USA, The Canaries, France and Spain. A tariff of 10 per cent of the market value is imposed by the UK on these imports but this does not deter sendings while prices are high.

CASE B. The exportation of apples from the USA to Brazil

The estimated national income per head of Brazil in 1968 was 263 US dollars whereas for the USA it was 3569 dollars according to the UN Year Book of Statistics, 1970. Although this case is concerned therefore with exportation of produce from a developed to a lesser developed country it does illustrate several important features in international trade.

Brazil as an importer of apples

Brazil produces apples, 13 000 metric tons in 1970 but most of them are used for processing. Even though it produces 95 per cent of home consumption of agricultural commodities it is not self sufficient in apples nor is it likely that more could be produced for climatic reasons. The Argentine is the traditional supplier of fresh apples but others are received from the USA, Canada, Uruguay, France and Greece. From September to April there are no southern hemisphere or Argentine apples in the markets and northern hemisphere apples are in demand.

Brazil has already doubled its population this century and is expected to do so again before the year 2000. It is the urban population that is growing fastest and with increasing incomes is an important group of consumers of apples. These consumers have shown a preference for quality Red Delicious apples.

Brazil is a member of the Latin American Free Trade Association (LAFTA) which imposes a 32 per cent *ad valorem* tariff (see page 121) on imports of apples. This is to protect members of the association chiefly the Argentine from competition. In effect it raised the cost of supplying fruit to Brazil by 32 per cent. Although Brazil is protected from the USA imports in this way it does have other strong associations with the USA through its membership of the Organisation of American States (OAS). This was formed in 1948 and has 24 members in the Caribbean, Central and South America. It aims to promote economic cooperation, social and cultural development and collective security for its members.

The USA as an exporter of apples

In contrast to Brazil the USA produced 2 880 000 metric tons of apples in 1970 and has exported for many years some consignments having been sent to Brazil. The USA is a vast country and the State of Ohio has been chosen as an example of an apple producing area. One reason for this choice is a recent bulletin from the Ohio Agricultural Research and Development Centre (Steele, H.L. and Vera-Filho, F.M. (1970). Economic feasibility of exporting apples to Brazil) which enables us to examine in greater detail the potential results of apple exportation.

Apples are the most valuable crop grown in Ohio but do not always sell well at home in the face of organised competition from other producers in the USA. In determining whether exportation is feasible the cost of production, packaging and transportation are important together with size of package, reliability of supply and the amount of promotion undertaken to stimulate demand. In the first instance apples from Ohio must be transported to a port, possibly New York. This is done in refrigerated trucks which is estimated to cost 55 cents per bushel. They are then shipped from New York to Rio de Janeiro this costing a further $2.50 per bushel including commission, insurance and freight (cif). On arrival at Rio de Janeiro a 10 per cent mark-up is likely by the wholesalers. The average cif price of US apples in Brazil was higher in 1966 and 1967 than prices at the same time in Argentina, despite the extra costs and mainly due to the time of supply but also because they were of superior quality.

In the Ohio feasibility study Cleveland, Ohio market was taken as a home standard with which to compare prices with those received from Rio de Janeiro. In January, February and March 1967, after deducting the Brazilian wholesale mark-up and all freight and transfer costs including the 32 per cent LAFTA tariff, the net prices to Ohio producers were 4, 91 and 74 cents above the Cleveland average prices respectively. For the remaining 9 months of the year net prices were less than the Cleveland average. If the tariff had not been in force the advantage over Cleveland would have increased to $2.05, $3.16 and $2.91 respectively. The tariff is clearly an effective barrier to trade.

CHAPTER 8

POLICY

A policy is a course of action adopted to bring about a desired change. In this chapter we are concerned with the workings of government policies towards horticulture. Individual growers and horticultural organisations may also have policies of their own to achieve particular ends.

National policies are usually directed towards some economic goal such as the expansion of output, export earnings, productivity, competitiveness, import saving or social considerations. International policies may arise from the joint action of countries having common problems or aims.

8.1 PRODUCER INCOMES

The potential of a crop to generate income for the grower will in most cases determine the quantity he will grow. If a crop becomes relatively more or less attractive to him, he will rationally expand or contract his production. In the same way by influencing the relative profitability of crops and as a result the grower's income, the volume of production can be manipulated by government policies. If growers' incomes are considered satisfactory or there is no national need for the crop or type of business, then government intervention may be unnecessary. All too often incomes are low in agriculture, both in absolute terms, and relative to other industries. We talk of the disparity of incomes between agriculture and other sectors of the economy. In practically all countries of the world the agricultural population receives less than the average earnings of the whole working population.

Low incomes can arise in any industry because of inadequate management. There are however certain characteristic features of agricultural production units that can lead to income problems, often needing remedial action by government policy. Some of these features will be recognised by the reader from earlier chapters and they are enumerated below.

Firstly surplus production, both in the short and long term, is a major cause of low incomes. In both cases supply exceeds demand and prices are depressed. In the short term these surpluses may commonly arise as the result of exceptionally good weather causing large yields. In strongly cyclic production situations the surpluses may accentuate the 'cobweb' effect. In the longer term, the impact of technology through improved cultivars, husbandry and extension, leads to larger yields and inevitably supply which further depresses prices. Technological advance increases supply; its benefits are bestowed on the producers who are first to use it.

A second major cause of low incomes in agriculture is associated with the general inelasticity of demand for agricultural products and the switch to 'convenience' consumption. As incomes increase consumers spend more on luxuries and less on food and as we have already seen this is to the benefit of the development of horticulture generally. However keeping pace with the evolution of the convenience market can raise difficulties for producers who must inevitably adopt new marketing techniques. The penalty for failing to keep abreast of these developments is a diminished share of the market. It is difficult for individual growers to enter the 'service' market; capital investment in packaging machines and the adoption of new techniques may not be possible without some government encouragement or assistance.

Thirdly, low incomes may stem from imbalances in resource use during economic development, particularly with respect to the use of capital and labour in relation to land. To ensure that adequate capital and labour of the right quality is employed policies may need to be implemented. There may well be a disparity in the general level of education between the rural and urban population of a country. This may be important in determining the rate of migration of workers, the receptivity of growers to extension services and their ability to interpret and make use of new technology. These considerations are not only relevant to developing countries; they are also important in the developed nations.

Fourthly, differences between regions within a country can cause income problems. Soil or climatic differences may result in disparities of potential production between regions and thus affect the incomes of the farmers and growers. This often leads to political unrest. Universally applied policies only maintain or may even accentuate the differences between the groups of producers. Where there is no alternative agronomic use for the land the problem may be particularly difficult to solve.

The reasons for government support of agriculture and horticulture

All countries of the world operate some form of policy towards agriculture and it is usual for this action to be aimed at providing for some national need or interest. The alleviation of the income problems discussed in the last section are an obvious reason for government intervention but we can identify other reasons.

Much of agricultural policy is achieved by transferring resources in the form of cash grants, subsidies and guaranteed prices to the agricultural sector. This is done rather than allow it to stand on its own feet and face competition which may result in a more efficient home industry but may also create the need for more importation of agricultural produce. This would be unacceptable to a developing country in the process of increasing exports. In developed countries too, a burden is placed on the balance of payments by being forced to increase imports of food. Indeed, at times of crisis with respect to the balance of payments it is common for governments of developed countries to deliberately encourage home production to save imports. Unfortunately this may cause a short term increase in cost as more resources are used. Trade agreements may further cloud an apparently simple situation. For example the export of industrial products to another country may necessitate the acceptance of imports of horticultural produce in exchange. The government in this case may have to decide on its priorities if the home horticultural industry is threatened by these imports.

A further complication may arise when agriculture itself becomes dependent upon imports. This is certainly true in European agriculture where much animal feed and all oils for glasshouse heating and machinery operation are imported.

Countries that have suffered or are threatened by war may well decide to maintain an efficient agriculture at home. They do this so that in the event of hostilities and perhaps enforced isolation they are able to feed themselves. Horticulture may have an important role to play in this strategy as the supplier of vitamins in the national diet. The effectiveness of such a policy depends upon how quickly agriculture can be expanded in emergency and how dependent the country is on imports of food. If the war is of the conventional type this policy may be worthwhile but a nuclear war would probably invalidate it.

The deliberate maintenance of a large rural population has been the aim of several countries. This was certainly true for example of the Physiocrate movement in France in the 16th and 17th centuries, whose members believed that all wealth was derived from the land. President Jefferson in the USA based his policies on 'owner-occupied, debt free and politically stable organisations,' and thus encouraged small-scale farmers. Large rural populations were also part of Nazi philosophy. Switzerland and Israel too, have deliberately written this into their national constitutions. The need for maintaining small farms to act as a step ladder for new entrants to farming is commonly voiced by farmers and growers.

Greater leisure time and mobility have brought a new dimension to the argument for giving support to agriculture. In the developed countries in particular, the countryside as an amenity is taking on a new importance. The appearance of the countryside is very much a function of the national policy towards agriculture and the income status of the farmer. A depressed agriculture with poor crops and run-down buildings is visually unattractive. At the other extreme it has to be admitted that some modern farming practices may also appear unsightly. It is argued that the cheapest way to maintain a pleasant countryside as an amenity is to give support to agriculture.

The reasons for supporting agriculture are diverse and the objectives of particular policies may gain strength according to who is advocating them. The economist concerned with the efficient use of resources may have no time for 'give-away' policies. The politician by contrast may be concerned with votes, particularly in rural areas, and when confronted by a strong political lobby may find it difficult to be truly objective in his views. The producer naturally seeks the highest price that he can get to maximise his income. Meanwhile the consumer seeks the lowest prices and expects continuity and variety of supply. Clearly all these views and some others are competitive. The relative strength of 'pull' may not be equal and opposite in the physical force sense but the resultant policy may well be a compromise policy based on the relative strengths of the advocates. A democratically determined policy may please all of them most of the time but none of them all the time. Regrettably perhaps, the general public are commonly ignorant not only of the objectives of a particular policy but also of the methods used to implement it.

8.2 THE MECHANISMS OF GOVERNMENT POLICIES

It is impossible to review all the different instruments of policy used in giving support to horticultural producers. The most commonly employed methods are described in terms of their theoretical base and particular advantages or disadvantages. Later in the chapter case studies are presented to illustrate some of them in operation.

1. Import restrictions

Only three types of import restrictions are considered here. Other methods are used and special licencing, total embargo, timetabling and even health restrictions have been discussed in the previous chapter.

(a) *The tariff* is a tax imposed on foreign suppliers collected at the port of entry or in the markets. If a particular amount is charged per unit it is called a *specific* tariff. If a percentage of the supply price is charged it is called an *ad valorem* tariff. The result is to raise the cost or lower the effective price to the foreign supplier, making the market less attractive to him. The sequence of events might be as follows

- the home government imposes a tariff of 10 per cent *ad valorem* on imports;
- the foreign supplier, recognising the lower price offered, contracts his supply;
- total supply at the home market falls and the price increases as a result of the reduction in foreign sendings;
- home producers recognise this higher price and expand their production accordingly.

The advantage to home producers is obvious and revenue from the tariff accrues to the government and may well be used as a source of income. However the tariff does not protect against technological advance or increased competitiveness on the part of the foreign supplier. The imposition of the tariff may give incentive to foreign suppliers to make these improvements so that they can defeat the tariff. Worse still, protected home producers may become complacent and neglect the maintenance of their own productivity, further weakening their position.

Foreign suppliers may actually retaliate positively by encouraging exporters with government aid in the form of direct financial assistance. This has the opposite effect of tariffs and is used by many countries under a variety of guises.

(b) *The quota* unlike the tariff is not revenue producing. It refers to a particular quantity or value of produce and is really a form of international supply control. By restricting the amount of produce supplied from overseas, prices are raised for home producers who can then expand their output. The establishment of a quota system may lead to a slowing down and even ossification of trading activity. Accusations are sometimes made regarding the honesty of the administrators of quota schemes suggesting that their allocation has become subject to bribery. Quotas are not acceptable to GATT and it is to be expected that they will disappear particularly in signatory countries.

(c) *The minimum import price and levy* is used in an attempt to balance the internal market prices and the external world prices, the national frontier being the boundary between them. In a common market group of countries the common external frontier is the boundary. Home market prices are maintained by setting a minimum price at which imports are accepted and by charging a variable levy. The levy is raised or lowered so that an external supplier cannot take advantage of higher prices in the home country. It has the same effect as the tariff in making the product effectively less competitive.

Usually the cost of transportation is taken into account. At certain times when prices are very low a total embargo may be imposed. The scheme may also be operated in reverse to finance exporters when prices are higher in the internal market. Funds accumulating from the levy may be retained by the government or diverted to other agricultural support schemes.

2. Market intervention schemes

(a) *Support buying* aims to maintain a minimum pre-defined price level. An agency provided with finance buys produce from the market when prices are below this level until the minimum price is re-established. The funds may be provided as part of government support for horticulture or be collected from growers as a levy on produce handled when prices are higher. Produce withdrawn from the market is either stored until the price improves, processed or destroyed.

The main problem associated with this mechanism is the setting of the price level at which support is given. If too high it will encourage more production which may lead to huge surpluses and a further need for support buying. If set too low production may be discouraged and political unrest may be instigated amongst growers. Sharply fluctuating yields from year to year only aggravate this problem.

For agricultural products that can be stored for very long periods such as cotton and hops this type of scheme may be operated on an international scale. These are called buffer stocks but they are not used in horticulture.

(b) *Supply control by marketing boards* has already been discussed in chapter 3. The boards may also be the agents for other support schemes, for example support buying, the administration of capital grants and sponsoring of research.

3. Capital assistance schemes

(a) *Direct grants* towards particular uses have the great advantage that moneys are directed at the point of real need. The most progressive and efficient of producers will be best able to organise themselves to use the money so that it goes to those who are perhaps most deserving. Sometimes only a proportion of the total cost of the project is made available and the user has to find the remainder from normal credit sources. Due to lack of knowledge of these and capital budgeting and investment appraisal some growers will have difficulty in making use of the funds. They are a direct drain on the reserves of the exchequer of a country and may be expensive if maintained beyond the time necessary to achieve the desired remedial effect in the industry. If used with the intention of increasing efficiency these schemes may also give incentive to expansion as a response to a favourable financial situation.

(b) *Low interest loans* make capital relatively low in price for growers compared with other sectors of the economy thus giving them incentive to invest further. Further stimulus may be given to guaranteeing the security of the loan. Special banks and sometimes cooperatives may be established to administer the funds. Expansion investment may be encouraged and even lead to over capitalisation if interest rates are made too attractive. The repayment of the loan may become a considerable burden to the grower particularly if it coincides with a difficult trading period of rising costs and low prices, the so called 'cost-price' squeeze.

(c) *Technological assistance* is given by most countries to agriculture and horticulture by fostering research and disseminating its findings through extension services. In some cases payment may be made for all or part of the service, in others it is given free. As with any free service there is the inherent danger that complacency will result.

This method of support is perhaps the oldest and although it inevitably increases supply it is necessary in all countries, particularly the developing nations where technological assistance is often the basis of aid from developed countries.

4. Structural improvement

The usual objective of this type of policy is to promote the development of more efficient production units. Typically, this means larger and fewer holdings using less labour and sometimes land than hitherto. There may be policies for example to prevent land being purchased for speculative gains particularly if it might jeopardise the chance of structural improvement. More specifically there may be schemes to make direct payments to individual producers to leave the land or to retire prematurely. Assistance towards the amalgamation of holdings may be given, often encountering severe social problems, but these have been successful in some countries where fragmentation of holdings has been a problem. Cooperation in both production and marketing may also be encouraged by technical and financial assistance.

8.3 CASE STUDIES OF HORTICULTURAL POLICY

A weakness of the use of case studies in economic analysis is that they become outdated. For example the accession of the UK to the EEC has changed the situations with respect to UK tariffs in Case A. The case study is still valuable however as an illustration of the mechanism in practice. The following cases are presented only to illustrate the use of particular implements of horticultural policy in reality. They are not intended as complete accounts of the current policies of the countries concerned• towards horticulture. The six cases represent each of the major categories from the last section.

Case A. The UK tariff on imports of tomatoes

A 10 per cent *ad valorem* (by value) tariff is imposed on imported tomatoes all the year round if the value of imports does not exceed 6p per 1 lb, or 5p from 16 to 31 May. From May to October there are further specific duties; this is the period in which the British crop is marketed. 1½p per lb is charged in the first half of May rising to 2½p per lb in the second half of May and the first half of June; 2p per lb is then charged until the end of July, 1½p for the whole of August and finally 1p per lb in September and October.

The greatest use of the tariff coincides with the time of most supply in August but between May and October approximately £20M worth of tomatoes are imported. We talk of the formal incidence of the tariff – the receipts from the tariff form a percentage of the total value of tomatoes. This usually exceeds 40 per cent in August when the main overseas supplier is Holland and it was to protect home growers from Dutch competition that the tariff was first introduced. Not only has it been in operation for 30 years but it gradually increased until 1961 to what is still the current level. During this time Dutch growers have made themselves very competitive in production and marketing to combat the tariff.

Case B. The EEC import regulations for fruit and vegetables

There has been a unified import policy for fruit and vegetables in the EEC since January 1970 and in keeping with GATT principles no quantitative restrictions are allowed. A common external tariff operates for some crops and is defined for specific time periods. Tomatoes for example are subject to an 11 per cent *ad valorem* tariff from November until mid May and 18 per cent for the remainder of the year.

The tariff alone has not been regarded as adequate protection by all member countries and additional safeguards have been introduced. The first of these is the so called reference price system, a form of minimum import price and levy. Reference prices are determined by the community from market prices prevailing over the previous three years in representative markets in the community. They are operative at the ports or frontiers so that if the price of imports is lower than the reference price the importers are obliged to pay a compensatory levy. The actual levy is variable and is the difference between the reference and import prices. Reference prices are applied to tomatoes, apples, pears, table grapes, plums, cherries, peaches and citrus fruit.

The second measure adopted by the EEC is to impose temporary bans or restrictions when imports from a non member country threaten to disturb the home market. This is essential because the reference price system offers no protection if produce is supplied to the EEC at prices bearing no relation to their true cost of production. This is called 'dumping'; the exporting country is satisfied to dispose of the produce in the knowledge that it will at least get some return for it. The use of anti-dumping duties is permissible within the rules of GATT if both dumping and damage to the home industry can be proved, which is not easy to do.

Case C. The Netherlands minimum price insurance scheme

In operation since 1948, this scheme has a 'support buying' function. It is of particular interest because support buying has become part of the EEC policy. The scheme is organised by the Central Bureau of Horticultural Auctions in the Netherlands. Most home grown produce passes through Horticultural Auctions. A minimum price is set and produce is bought from the market whenever the price falls below this level. To accumulate an adequate fund for this purpose a levy is collected from growers whenever the price of their produce is higher than the minimum. The fund also finances the disposal of produce withdrawn from the market. This may be costly because the produce withdrawn is likely to be inferior but it helps the maintenance of high quality standards. Each commodity has a separate fund which prevents a situation where, for example, cucumber growers feel that they are subsidising tomato growers. There were 37 product lines in 1970.

The support prices are not determined in advance so they are not a forward guarantee to producers but the setting of the minimum price is a crucial feature. If set too high over-production is encouraged and this has been a major problem of EEC policies. If set too low it does not fulfil its aim of price support. Grading standards must also be high. In practice the schemes are appraised every year and changes are recommended each time; so far agreement to continue has been reached. Since 1967 the central fund has taken on new functions and now provides publicity, research into containers and chemical residue problems. In 1969 27M guilders were collected and 20.8M guilders paid out as compensation for surplus. The income to the fund from interest on the balance was considerable and some schemes can now operate free from charge.

Case D. Israel: marketing boards and export company

Israeli horticultural production is planned and controlled through the vegetable and fruit production marketing boards comprised of growers, wholesalers, consumers and Government officers.

The Fruit Board coordinates fruit tree raising and orchard planting according to national plans. Licences are issued for the transportation and sale of fruit in the wholesale markets. Levies are collected from the licences to provide funds. Surpluses may be disposed of by the Board to processing industries, the army, new settlements or welfare institutions. The Board is also a part owner of the Export Marketing Company and undertakes the supply of fruit for export.

The Vegetable Board controls the production and supply of vegetables throughout the country by the supervision of production quotas and by assuring minimum prices to growers. Grading and packing standards, storage, transportation and research are also part of the Board's function. A crop insurance scheme is also operated. As with the Fruit Board surpluses are disposed of and the Board owns part of the Export Company.

The Export Company sends mainly to Europe which is a natural nearby market for Israeli horticultural produce, exotic and out of season crops being very acceptable in these markets. In 1971 the UK received 30 to 40 per cent of these exports.

The Agricultural Export Marketing Company (Agrexco) was established and is owned jointly by the marketing and production Boards and the Government. It handles all horticultural produce for export with the exception of citrus. 'Carmel' is the brand name of the company under which all exports are sold. Regional offices in each important recipient country further the expansion of particular markets and entry to new ones. All exports are inspected by the Ministry of Agriculture to standards set in accordance with those of the EEC and the grower pays for the service. Agrexco acts as transport agent to the ports, collecting the produce from the grower.

Case E. The UK Horticultural Improvement Scheme (HIS)

Since 1960 direct grants have been available to horticultural producers in the UK with the aim of increasing competitive efficiency in the industry. In particular they are intended to prepare growers for the day when tariff protection is reduced through the activities of GATT and general trade liberalisation. The scheme has since been modified but basically grants are made amounting to a third of the total capital cost of re-equipping and modernising holdings with particular emphasis placed on marketing facilities. Growers' projects have to conform to minimum standards and are administered and implemented through the development and advisory service ADAS. To qualify, the grower must farm at least four acres of land or its equivalent in terms of potential output of glasshouse, 'forcing shed' or watercress bed area. Furthermore the grower must have farmed the land for at least two years and the whole unit must be capable of yielding him an adequate income. Grants are also available for market renovation and rebuilding, the furthering of cooperation and for grubbing orchards.

When the scheme started in 1960 £8M was made available and the scope for its use was restricted. By 1963 only £4½M had been spent and most of this was used to develop fruit cooperatives and storage facilities. Although there were 70 000 holdings qualifying for grants only 5300 projects were sanctioned. In 1964 the scope of the scheme was broadened and a further £50M was made available until 1974. Some grants were increased in 1966 to 38½ per cent. The result is that over 60 000 projects have now been approved and the grants have been more evenly shared between the different sectors of horticulture.

Glasshouse growers in particular, often regarded as the most vulnerable to competition, used 60 per cent of the grants between 1965 and 1967. The financial incentive has stimulated growers to make expansion investments with the result that the national glasshouse area has increased. This has commonly been through the erection of unheated glass to which heating can still be installed at some later date. Expansion investments of this type take time to become productive so that they do little to raise the efficiency of the industry in the short term.

Case F. Structural reform policies in the EEC

A large part of EEC policy is aimed at market support but in recent years this has led to over supply of some products. In addition farms are often small and fragmented, the average size being only 11 hectares but they are increasing slowly. There are considerable regional problems within the community. Most member countries had schemes for structural reform long before the inception of the EEC.

In 1968 a comprehensive plan was drawn up for the reformation of agriculture including horticulture. Freedom of choice, flexibility and national individualism were the key factors in this document. The community anticipated payment of half the cost of the scheme but individual Governments were to bear the responsibility for its implementation at home. It was agreed that farmers should not be forced to accept reforms and the reformers were also to take account of the diversity of potential for production and social aspects of the different parts of the community.

The main aim of the new policies was to reduce the land area devoted to agriculture and to reorganise the structure of farm units. The latter may be antagonistic to the former in that improving farm business structure may only lead to further increase in output adding to the existing over-production of come crops. To encourage the withdrawal of land from agriculture, subsidies, tax concessions and grants were proposed for owners to help them establish afforestation schemes, parks, green belts and other recreational facilities.

Structural reform of farm units is proposed by actively encouraging an outflow of surplus manpower from agriculture. Eight times the rental value of land has been offered as a surrender grant, payable to owner occupiers giving up farming. The farm would then be amalgamated with another farm or put to another use. Training schemes are envisaged for owners and their families to enable them to settle in new occupations. Knowledge of alternative employment is made available through information offices. Premature retirement inducements are made to farmers over 55 years of age, the amount and the pension received depend upon their circumstances but migration from agriculture is a prerequisite of their receipt.

Making farm units more viable necessitates reorganising them so that they produce adequate incomes. Viability may be achieved by total amalgamation of two or more holdings but it can be done on a partial basis. Agreement may be reached between owners to share a particular facility but to retain their individual business identity. Inevitably financial, legal and social problems arise from any form of amalgamation. The community makes available investment grants, credit facilities and training schemes directed towards improving business viability.

Prices are a crucial feature in determining producers' incomes and in the EEC guaranteed prices have encouraged over-production. Low level price guarantees would clearly hasten both the exodus from farming and the desire to restructure to become more efficient. The dilemma facing the EEC policy makers is that there is a great disparity between regions within the community with respect to income potential for producers. It is only the higher price guarantees that maintain an income for some of these producers.

APPENDIX 1

HORTICULTURAL MARKETING

THE FLOW OF HORTICULTURAL PRODUCE PRODUCER TO CONSUMER

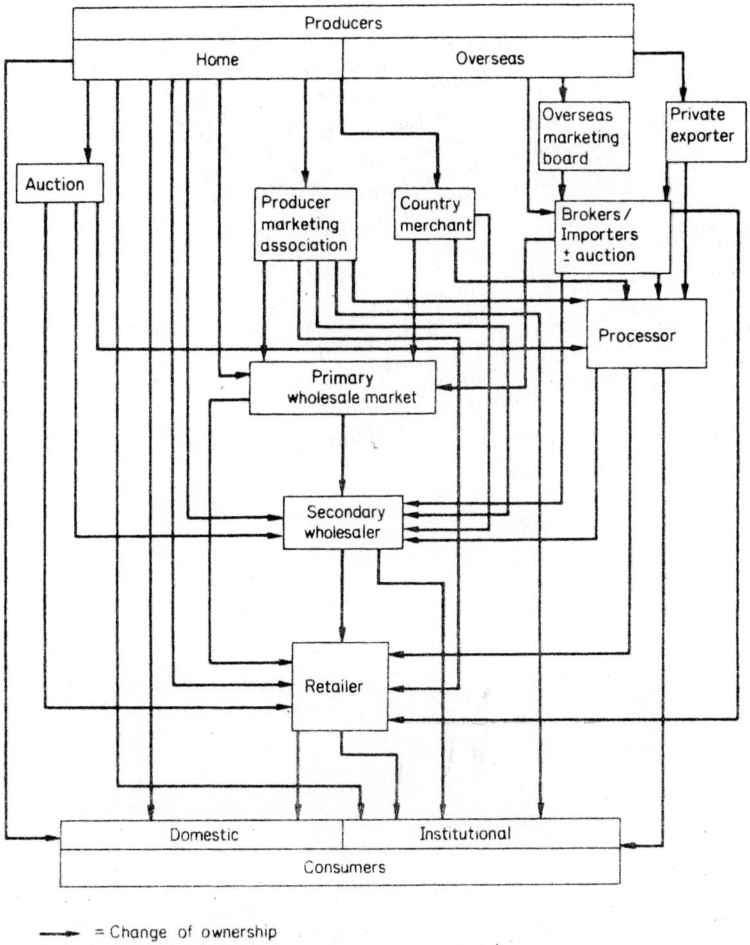

APPENDIX 2

SOME USEFUL CONVERSIONS

Length	1 mile	=	1.609 kilometres
	1 yard	=	0.9144 metres
	1 foot	=	0.3048 metres
Area	1 square mile	=	2.590 square kilometres
	1 acre	=	0.4047 hectares
	1 square yard	=	0.8361 square metres
Weight	1 Imperial ton	=	1.016 metric tonnes (1000 kilogrammes)
	1 hundredweight	=	50.80 kilogrammes
	1 pound	=	0.453 kilogrammes
Volume	1 cubic yard	=	0.7646 cubic metres
	1 cubic foot	=	0.0283 cubic metres
	1 Imperial gallon	=	4.546 litres
*Currency	1 pound sterling (£)	=	2.49 US Dollars
	(May 1973)	=	15.45 Danish Kroner
			11.00 French Francs
			7.05 West German Deutsche Marks
			661 Japanese Yen
			7.31 Netherlands Guilders
			8.06 Swiss Francs
			1.76 Australian Dollars
			1.88 New Zealand Dollars
			15.02 Brazilian Cruzeiros

* *At the time of publication exchange rates are subject to wide variation; the rates quoted should be checked before applying them to practical uses.*

FURTHER READING

1. *General economics*

CAIRNCROSS, A. (1966). *Introduction to economics*, Butterworth, London.

CARTER, C. F. (1967). *The science of wealth*, Edward Arnold, London.

DEANE, P. and COLE, W. A. (1969). *British economic growth 1688-1959*, Cambridge University Press.

LIPSEY, R. G. (1966). *Introduction to positive economics*, Weidenfeld and Nicolson, London.

NEEDHAM, D. (1969). *Economic analysis and industrial structure*, Holt Rinehart and Winston, New York.

PAISH, F. W. (1964). *Benham's economics*, seventh edition, Pitman, London.

SANDFORD, C. T. and BRADBURY, M. (Eds) (1970). *Case studies in economics* (2 vols), Macmillan, London.

2. *Agricultural and horticultural economics*

ALLEN, G. R. (1959). *Agricultural marketing policies*, Blackwell, Oxford.

BISHOP, C. E. and TOUSSAINT, W. D. (1963). *An introduction to agricultural economic analysis*, Wiley, New York.

CAPSTICK, M. (1970). *The economics of agriculture*, George Allen and Unwin, London.

CLARK, C. and HASWELL, M. (1970). *The economics of subsistence agriculture*, fourth edition, Macmillan, London.

COWLING, K., METCALF, D. and RAYNER, A. J. (1970). *Resource structure of agriculture: an economic analysis*, Pergamon Press, Oxford.

EICHER, C. and WITT, L. W. (Eds) (1964). *Agriculture in economic development*, McGraw Hill, New York.

ELLIS, P. G. (1970). *Horticulture in Britain*, Part 1, *Vegetables*, Part 2, *Fruit and flowers*, Ministry of Agriculture, Fisheries and Food, HMSO, London.

HALLETT, G. (1969). *Economics of agricultural policy*, Blackwell, London.

HEADY, E. O. and JENSEN, H. R. (1961). *Farm management economics*, Prentice-Hall, New York.

METCALF, D. (1969). *The economics of agriculture*, Penguin Books, Harmondsworth, England.

McCRONE, G. (1962). *The economics of subsidising agriculture*, George Allen and Unwin, London.

SARGENT, E. D. and ROGERS, S. J. (Eds) (1971). *The economic prospects for horticulture*, Oliver and Boyd, Edinburgh.

SELF, P. and STORING, H. (1962). *The state and the farmer*, George Allen and Unwin, London.

SHEPHERD, G. S. and FUTRELL, G. A. (1969). *Marketing farm products*, Iowa State University Press.

TRACY, M. (1964). *Agriculture in Western Europe*, Jonathan Cape, London.

INDEX